TEXT BOOK
OF
ELECTROSTATICS

DPH PHYSICS SERIES

TEXT BOOK OF ELECTROSTATICS

By

D.K. Jha

Reprinted - 2019

First Published - 2005

ISBN: 978-81-8356-027-6

Text Book of Electrostatistics

Published by:
DISCOVERY PUBLISHING HOUSE PVT. LTD.
4383/4B, Ansari Road, Darya Ganj
New Delhi-110 002 (India)
Phone: +91-11-23279245, 23253475; 43596065
E-mail: discoverybooksindia@gmail.com
discoverypublishinghouse@gmail.com
web: www.discoverypublishinggroup.com

Printed at:
Infinity Imaging Systems
Delhi

Preface

Text book of electrostatics has been designed to cover the syllabus of Electrostatic for B.Sc. students of the Indian Universities. The subject matter has been arranged so as to provide a clear and integrated approach to the subject with all essential tools of applicable mathematics required for B.Sc. curriculum. Illustrated examples have been incorporated to help the students in getting the clear concept to the subject and allied matter.

Care has been taken to make the treatment of the subject simple and accessible to the average students. It believed that the book in to present from will be found to be useful by the students community and the teaching traternity alike.

Suggestion for the improvement of the book will always be most welcome.

D.K. Jha

Contents

Alternating Currents

DESCRIPTION OF AN ALTERNATING ELECTRIC CURRENT

An alternating current, in general, is a current that is a periodic function of time. In other words, an *alternating current is one which passes through a cycle of changes at regular intervals.* Each cycle consists of two half cycles during one of which the current is entirely positive whereas during the following half-cycle the current is entirely negative.

In an alternating current circuit we are concerned with a steady state current and voltage which are oscillating sinusoidally without change in amplitude.

The variation of emf is written as,

$\xi = \xi_0 \sin \omega t$ and the corresponding current is

$$I = I_0 \sin \omega t$$

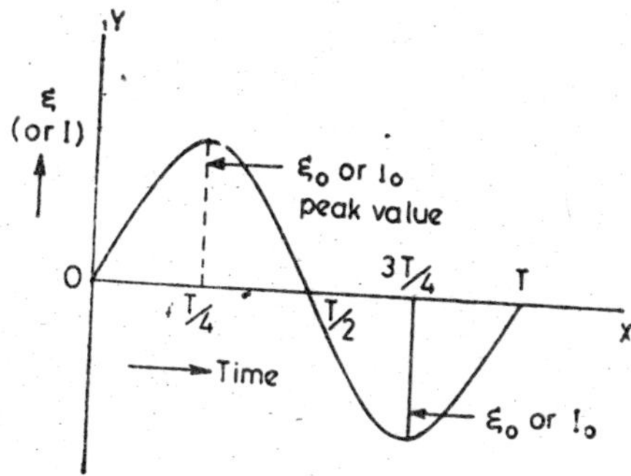

Fig. 1.1

where ξ_o is the amplitude or peak value of voltage ξ represents the voltage at any time t and $\omega = 2\pi f$ is the angular frequency of the voltage applied. The periodic time of the oscillation is $T = 2\pi/\omega$. The term ωt is called the phase angle. The emf and the current first rise to maximum in one direction and fall to zero. The direction then quickly reverses so that emf and the current rise to maximum in the reverse direction and again fall to zero. This completes one cycle of A. C. voltage or current as shown in Fig. 1.1. *The number of such cycles per second is called the frequency (f).*

AVERAGE OR MEAN VALUE OF ALTERNATING CURRENT OR VOLTAGE WAVE OVER ONE COMPLETE CYCLE

The value of current at any instant t is given by,

$$I = I_o \sin \omega t$$

The average value of a sinusoidal wave over one complete cycle is given by,

$$I_{av} = \frac{\int_0^T I_o \sin \omega t \, dt}{\int_0^T dt} = -\frac{I_o}{\omega} \frac{\left|\cos \omega t\right|_C^T}{T}$$

$$= -\frac{I_o}{\omega T}\left|\cos\frac{2\pi t}{T}\right|_0^T, \text{ (since } \omega = 2\pi/T)$$

$$= -\frac{I_o}{\omega T}[\cos 2\pi - \cos 0] = -\frac{I_o}{\omega T}[1-1] = 0. \quad ...(1.1)$$

Thus, *the average value of alternating current over one complete cycle is zero. The same is also true for alternating emf.*

AVERAGE VALUE OF ALTERNATING CURRENT OR VOLTAGE DURING HALF CYCLE

However, the average over one half-cycle is finite. It can be obtained very easily as follows:

$$I_{av} = \int_0^{T/2} \frac{I_o \sin \omega t \, dt}{\int_0^{T/2} dt}$$

$$= -\frac{I_o}{\omega \frac{T}{2}}[\cos\omega t]_0^{T/2}$$

$$= -\frac{I_o}{\frac{2\pi}{T}.\frac{T}{2}}\left[\cos\frac{2\pi}{T}t\right]_0^{T/2}$$

$$= -\frac{I_o}{\pi}[\cos\pi - \cos 0] \qquad ...(1.2)$$

$$= 0.637\ I_o \qquad ...(1.3)$$

Similarly, for sinusoidal alternative voltage, average value over one half cycle is,

$$\xi_{av} = \frac{2\xi_o}{\pi} = 0.637\ \xi_o \qquad ...(1.4)$$

The average value during the second half cycle is given by,

$$I_{av} = \frac{\int_{T/2}^{T} I_o \sin\omega t\, dt}{\int_{T/2}^{T} dt}$$

After evaluating the above equation,

$$I_{av} = -\frac{2I_o}{\pi} = -\ 0.637\ I_o \qquad ...(1.5)$$

The average value of alternating current (or voltage) during the first and second half cycles are equal but opposite in sign i.e., they are alternately positive and negative so that the average over one cycle is zero.

ROOT MEAN SQUARE OR VIRTUAL OR EFFECTIVE VALUE OF SINUSOIDAL VOLTAGE OR CURRENT

The effective or virtual value of an alternating current is that steady current which, when passed through a given resistance for a certain time will develop the same amount of heat as the as the actual alternating current shall develop when passed for the same time.

The rate of heat generation in a current carrying resistance is given by I^2T. As we are interested in the average or mean value of this quantity, which we shall write as $\overline{I^2}$ where $\overline{I^2}$ represents the mean value of I^2. Thus,

$$I_{eff}^2 = \overline{I^2} \quad \text{or} \quad I_{eff} = \sqrt{\overline{I^2}} \qquad ...(1.6)$$

Thus, the effective value of an alternating current is the square root of the mean of the squares of the instantaneous values over one complete cycle. Due to this reason the effective current is also called the root main square (RMS) current.

Let us calculate the average of I^2. If the alternating current is represented by,

$$I = I_o \sin \omega t$$

$$I^2 = \frac{\int_0^T I^2 dt}{\int_0^T dt} = \frac{\int_0^T I_o^2 \sin^2 \omega t dt}{T}$$

$$= \frac{I_o^2}{2T}\int_0^T (1-\cos 2\omega t)\,dt$$

$$= \frac{I_o^2}{2T}\left[t - \frac{\sin 2\omega t}{2\omega}\right]_0^T$$

$$= \frac{I_o^2}{2}$$

$$\text{Hence, } I_{eff} = \sqrt{\frac{I_o^2}{2}} = \frac{I_o}{\sqrt{2}} = 0.707\ I_o \qquad ...(1.7)$$

Thus, the effective or RMS value of an alternating current is $1/\sqrt{2}$ *times its peak value or maximum value. Similarly, the effective or virtual value of alternating emf is given by,*

$$\xi_{eff} = \xi_o/\sqrt{2} = 0.707\ \xi_o \qquad ...(1.8)$$

Form Factor

It is defined as the ratio of the virtual or R.M.S. value to the average value of an alternating current or voltage. Thus,

$$\text{Form factor} = \frac{I_{rms}}{I_{av}} = \frac{\xi_{ms}}{\xi_{av}} \qquad \text{...(1.9)}$$

Substituting values from eqs. (1.2) and (1.7), we get

$$\frac{I_o}{\sqrt{2}}\frac{\pi}{2I_o} = \frac{\pi}{2\sqrt{2}} = 1.11 \qquad \text{..(1.10)}$$

The form factor gives an indication of the wave shape of an alternating voltage or current.

A more general form of alternating current or voltage may be written as,

$$\xi = \xi_o \sin(\omega t + \propto) \qquad \text{...(1.11)}$$

and the current $I = I_o \sin(\propto t + \propto)$

where $\propto$ is called phase angle and added to the phase ωt. This addition does not affect the virtual or effective value (ξ_{eff}). ξ_{eff} is still given by eq. (1.8) which you can verify.

A.C. CIRCUITS

We considered the transient currents in a circuit after a switch is opened or closed. The problem of determining the steady state condition was relatively simple since the circuits were energized by direct current sources (batteries). The currents after a long time were either zero or else could easily be calculated by means of Ohm's law.

An interesting and perhaps more important problem is to determine what happens when an alternating current generator is used in place of battery. Just as in the direct current case, we have to solve the differential equation for a particular circuit. The solution of the differential equation for a particular circuit may be written as the sum of complimentary function and a particular solution. The complementary function which represents the transient solution has the same form when an alternating current source is used as when a battery is used. The complementary function may be expressed in terms of exponential functions of time. For all circuits having some resistance, the exponential terms have negative real parts. Consequently, after a relatively short time the value of the transient (current or voltage) is practically zero.

The new problem in the alternating current case is to determine the steady-state solution, which is, of course, the particular solution of the differential equation. For example, the circuit in Fig. 1.2 shows an alternating current generator connected to a resistance R and a capacitance C in series. Here, *the steady state current is not zero as in the case of direct current. In one cycle, this capacitor is charged, discharges, recharged oppositely, and then discharged once more.* This happens again and again.

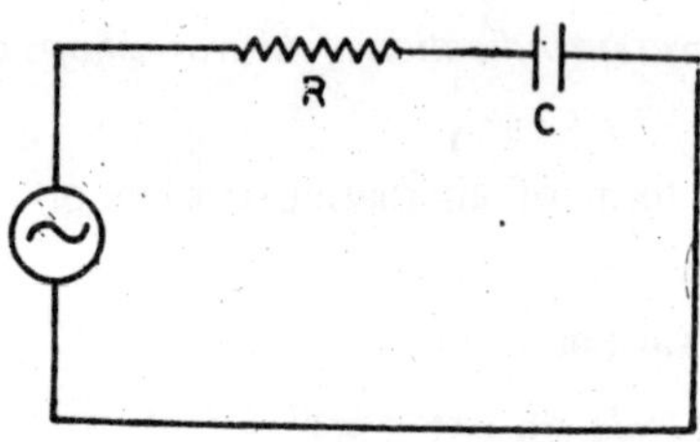

Fig. 1.2

The objective of this Chapter is to determine the current in such circuit. Basically there are two methods, namely,

1. The vector method; and
2. The complex number method.

We shall develop the complex number method.

CIRCUIT ANALYSIS USING COMPLEX NUMBER REPRESENTATION

The A.C. network analysis becomes *simple and convenient by the complex number representation. A combination of real and imaginary quantity is called a complex quantity.* A complex number can be written as,

$$\vec{Z} = x + jy \qquad ...(1.12)$$

where $\vec{Z}$ is a complex number having a real part x and an imaginary part jy, where $j = \sqrt{-1}$. Equation (1.12) expresses the complex number in algebraic (or rectangular form).

The complex numbers are represented in Cartesian coordinates by a vector in a complex plane by choosing real axis along the x-axis and imaginary axis along the y-axis as shown in Fig. 1.3

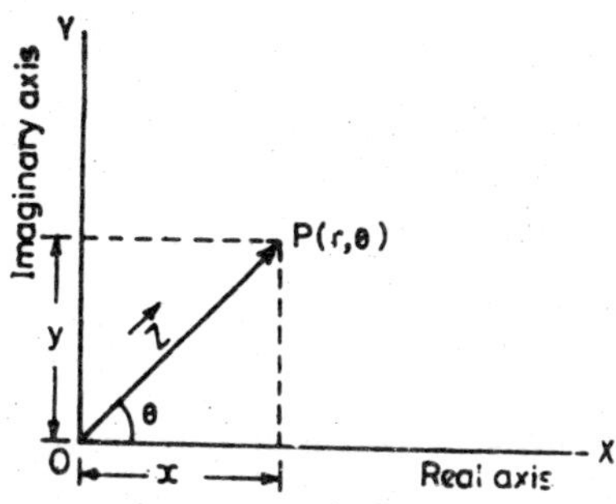

Fig. 1.3

The complex number is completely defined by point P since the projection along the real axis of line $\vec{OP}$ gives the real part, and its projection along y-axis gives the imaginary part of the number.

The length of the line ($\vec{OP} = \vec{Z}$) represents the absolute value or modulus (magnitude) of the complex number and angle θ is the phase of this number. Thus, the vector properties of complex quantities suggest that it may be convenient to express them in terms amplitude and direction as well as in real and imaginary parts. Hence, the complex number can be expressed in the following trigonometrical (polar) form as seen from Fig. 1.3. If point P has polar coordinates (r, θ) then we have,

$$x = r\cos\theta; \quad y = r\sin\theta$$

After substituting in Eq. (1.12), we get,

$$\vec{Z} = r\cos\theta + j\,r\sin\theta$$

$$= r(\cos\theta + j\sin\theta)$$

$$= r\,e^{j\theta} \qquad ...(1.13)$$

where $r = \sqrt{x^2 + y^2}$ ...(1.14)

and $\theta = \tan^{-1}(y/x)$ ...(1.15)

Important Points About Complex Numbers

1. Equality

Two complex numbers are said to be equal if both their real parts are equal and their imaginary parts are equal. Thus if,

$$Z_1 = x_1 + jy_1$$

$$Z_2 = x_2 + j\,y_2$$

then the equality, $Z_1 = Z_2$

gives the two real relations,

$$x_1 = x_2 \text{ and } y_1 = y_2$$

2. Complex Conjugate

We define the complex conjugate to $Z_1 = x_1 + j\,y_1$ as,

$$\overset{*}{Z_1} = x_1 - jy_1 \qquad \text{...(1.16)}$$

$\overset{*}{Z_1}$ is a complex number such that its product with Z_1 is.

$$Z_1\,\overset{*}{Z_1} = x_1^2 + y_1^2$$

3. Addition and Subtraction

Complex numbers are added and subtracted like vector quantities. Let the complex numbers be represented by,

$$\vec{Z}_1 = x_1 + j\,y_1$$

and $$\vec{Z}_2 = x_2 + j\,y_2$$

Then the resultant complex number will be (Fig. 1.4)

$$\vec{Z}_{Ra} = \vec{Z}_1 + \vec{Z}_2 = (x_1 + x_2) + j\,(y_1 + y_2) \text{ addition} \qquad \text{...(1.17)}$$

and $$ZR_s = Z_1 - Z_2\,(x_1 - x_2) + (y_1 - y_2) \text{ subtraction} \qquad \text{...(1.19)}$$

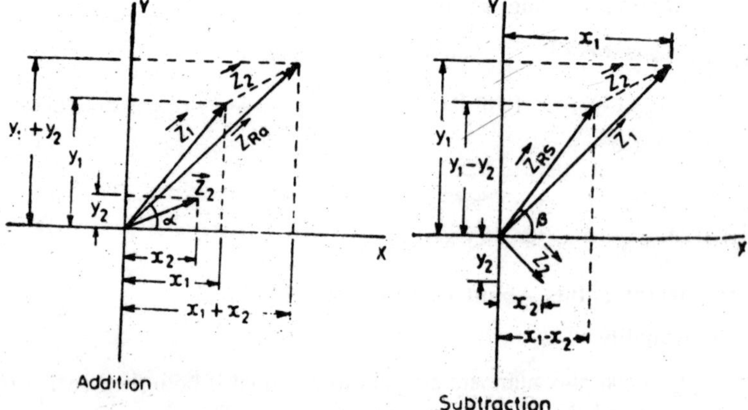

Fig. 1.4

The magnitude of the resultant complex number is given by

$$\left.\begin{aligned} ZR_a &= \sqrt{(x_1 + x_2)^2 + (y_1 + y_2)^2} \text{ addition} \\ ZR_8 &= \sqrt{(x_1 - x_2)^2 + (y_1 - y_2)^2} \text{ substraction} \end{aligned}\right\} \quad ...(1.19)$$

and the phase by

$$\left.\begin{aligned} \propto &= \tan^{-1} \frac{y_1 + y_2}{x_1 + x_2} \text{ addition} \\ \text{and} \quad \beta &= \tan^{-1} \frac{y_1 - y_2}{x_1 - x_2} \text{ subtraction} \end{aligned}\right\} \quad ...(1.20)$$

We can generalise the above rule that the sum of several complex number is a complex number whose real part is the algebraic sum of the real part of the added numbers and its imaginary part is the algebraic sum of the imaginary parts of the added numbers. The same is also true for subtraction.

4. Multiplication and Division

Products of complex numbers are defined according to the usual algebraic rules. For example, the product of two complex numbers Z_1 and Z_2 if expressed by Z_m is given by,

$$\begin{aligned} \vec{Z}_m &= \vec{Z}_1 \vec{Z}_2 = (x_1 + j\, y_1)\,(x_2 + j\, y_2) \\ &= (x_1\, x_2 - y_1\, y_2) + j\,(x_1\, y_2 + y_1\, x_2) \end{aligned} \quad ...(1.21)$$

It can be seen that sums and products of complex numbers are also complex numbers.

Division

Division can be defined as the inverse of multiplication. When complex numbers are divided, it is first necessary to eliminate the imaginary unit contained in the denominator and then to divide the complex numbers termwise.

$$\text{Thus, } \vec{Z}_d = \frac{\vec{Z}_1}{\vec{Z}_2} = \frac{x_1 + jy_1}{x_2 + jy_2} \quad ...(1.22)$$

To eliminate the imaginary unit contained in the denominator, multiply the numerator and denominator by the complex conjugate of the denominator.

$$\vec{Z}_d = \frac{\vec{Z}_1}{\vec{Z}_2} \cdot \frac{\vec{Z}_2^*}{\vec{Z}_2^*} = \frac{x_1 + jy_1}{x_2 + jy_2} \cdot \frac{x_2 - jy_2}{x_2 - jy_2}$$

$$= \frac{x_1 x_2 + y_1 y_2 + j(x_2 y_1 - x_1 y_2)}{x_2^2 + y_2^2}$$

$$= \frac{x_1 x_2 + y_1 y_2}{x_2^2 + y_2^2} + j\frac{x_2 y_1 - x_1 y_2}{x_2^2 + y_2^2} \qquad ...(1.23)$$

However, *the complex numbers can be multiplied and divided conveniently with less effort by transforming them from algebraic form to polar form (i.e., exponential form). By equation (1.13),*

$$\vec{Z}_d = r_1 e^{j\theta_1} \text{ and } \vec{Z}_2 = r^2 e^{j\theta_2}$$

Then, $$\vec{Z}_m = \vec{Z}_1, \vec{Z}_2 = r_1 r_2 e^{j(\theta_1 + \theta_2)} \qquad ...(1.24)$$

and $$Z_d = \frac{r_1}{r_2} e^{j(\theta_1 - \theta_2)} \qquad ...(1.25)$$

Thus, the product of two complex numbers is a complex number whose magnitude is the product of the magnitudes of the given complex numbers and the phase is the algebraic sum of the cofactors. The division (quotient) of two complex numbers is a complex number, the magnitude of which is the quotient of the magnitudes of the dividend and divisor, while its phase is the algebraic deference of the phases of the dividend divisor. Fig. 1.5 shows the magnitudes and phases of the two complex numbers after multiplication and division.

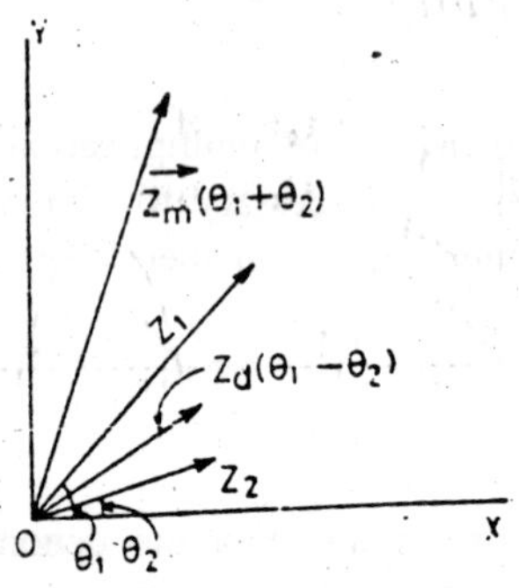

Fig. 1.5

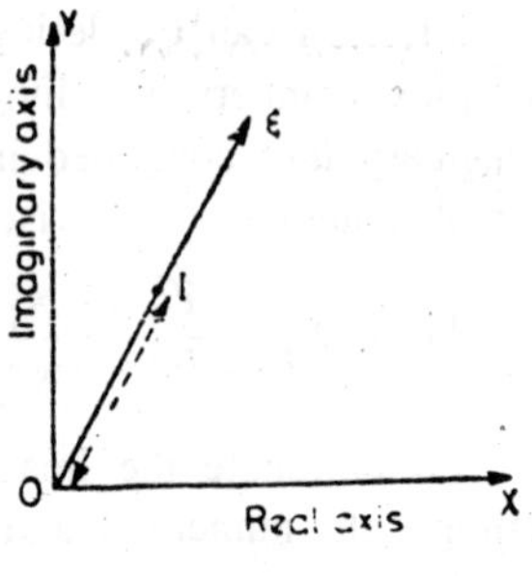

Fig. 1.6

COMPLEX REPRESENTATION OF A.C. CIRCUIT QUANTITIES

We know that any vector in a complex plane can be completely specified by complex number. This suggests that the A.C. network analysis can be simplified by expressing sinusoidal voltages and currents in complex number representation rather than in vector form. *Hence the sinusoidal voltage $\xi = \xi_0 \sin(\omega t + \alpha)$ and current $I = I_0 \sin(\omega t + \alpha)$ as expressed in eq. (1.11) that is general forms are nothing but the imaginary parts of $\xi_0 e^{j(\omega t + \alpha)}$ and $I_0 e^{j(\omega t + \alpha)}$ respectively.* Thus, alternating voltage and current can e represented conveniently by complex expression. The complex representation of voltage and current is shown in Fig. 1.6.

CIRCUIT CONTAINING PURE RESISTANCE ONLY

Fig. 1.7 (a) shows a circuit containing a pure resistance R with an alternating voltage $\xi = \xi_0 \sin \omega t$ applied across it. This voltage is the imaginary part of the complex number $\xi_0 e^{j\omega t}$. Thus, in complex number representation we can write,

$$\xi = \xi_0 e^{j\omega t} \qquad ...(1.26)$$

If I is the instantaneous current in the circuit, at any time t, then emf equation of the circuit is,

$$RI = \xi_0 e^{j\omega t} \qquad ...(1.27)$$

or $$I = \frac{\xi_0 e^{j\omega t}}{R} \quad \text{or} \quad I = I_0 e^{j\omega t} \qquad ...(1.28)$$

where $I_0 = \xi_0/R$

is the *maximum value of the current.*

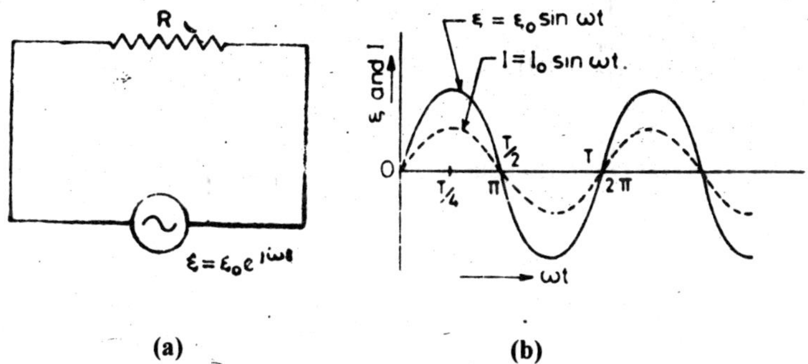

(a) (b)

Fig. 1.7

Equation (1.28) represents the variation of current with time t in the circuit. Comparison of Eqs. (1.28) and (1.26) shows that the current has the same form and frequency as the applied voltage. *Thus the current is in phase with applied voltage as shown in* Fig. 1.7 (c) shows the complex number representation of voltage and current.

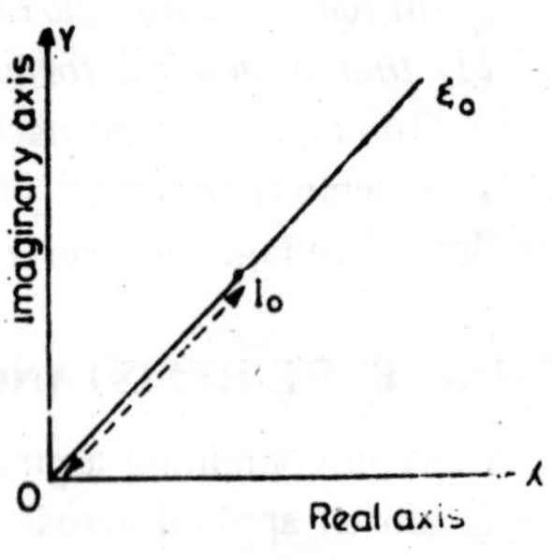

Fig. 1.7(c)

Here R is the impedance of the resistance, which when multiplied by the current through the element gives potential difference across that element. *The inverse of the impedance denoted by Y is usually called admittance.*

Thus,

$$\text{Admittance,} = Y = \frac{1}{Z} = \frac{1}{R} \qquad \text{...(1.29)}$$

CIRCUIT CONTAINING PURE INDUCTANCE ONLY

Fig. 1.8 (a) shows a circuit containing an inductor of self-inductance L with an alternating voltage $\xi = \xi_o \sin \omega t$ applied across it. This voltage is the imaginary part of the complex number $\xi_o e^{j\omega t}$. Thus, in complex number representation, we can write,

$$\xi = \xi_o e^{j\omega t}$$

Let I be the instantaneous current in the circuit, Further let dI/dt be the rate of change of current at any instant. Then emf induced in the inductor is given by,

$$\xi_{ind} = -L dI/dt$$

The negative sign indicates that the induced emf opposes the change of current. To maintain the current flow through the inductor, the applied voltage must be equal and opposite to the induced voltage. Thus, the emf equation of the circuit is,

$$\xi = -\xi_{ind} = L\ dI/dt$$

or $$\xi - LdI/dt = 0$$

or $$\xi_0\ e^{j\omega t} - L\ dI/dt = 0$$

or $$dI/dt = \frac{\xi_0 e^{j\omega t}}{L}$$

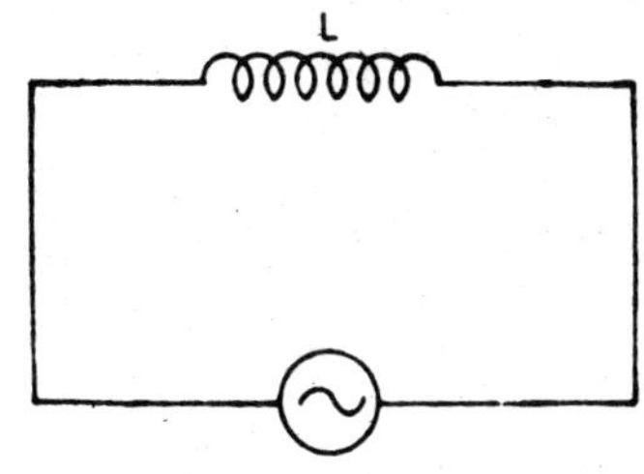

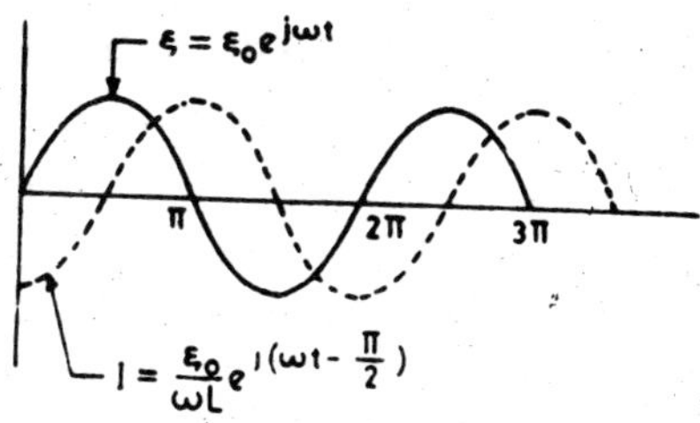

(a) (b)

Fig. 1.8

On integration, we get,

$$\int dI = \int \frac{\xi_0 e^{j\omega t}}{L} dt$$

$$I = \frac{\xi_0}{L} - \frac{1}{j\omega} e^{j\omega t} \qquad ...(1.30)$$

Now, $\frac{1}{j} = -j = e^{-j\pi/2}$

[since $e^{-j\pi/2} = \cos \pi/2 - j \sin \pi/2 = 0 - j = -j$]

therefore,

$$I = \frac{\xi_0}{\omega L} e^{j(\omega t - \pi/2)}$$

The quantity ω L is called *inductive reactance* of the inductance and denoted by X_L. Thus

$$X_L = \omega L$$

Hence,

$$I = \frac{\xi_o}{X_L} e^{j(\omega t - \pi/2)} = I_o \, e^{j(\omega t - \pi/2)} \qquad ...(1.31)$$

where $I_o = \xi_o/X_L$ is the peak value of the current.

Thus, $$X_L = \omega_L = \frac{\xi_o}{I_o} \qquad ...(1.32)$$

Thus, the inductive reactance is obtained by dividing peak value of voltage by peak value of current. Again, $X_L = \omega_L = 2\pi f_L$ shows that the *inductive reactance is directly proportional to the frequency of the A.C. supply voltage. Its value for D.C. supply is zero.* Further, since,

$$X_L = \frac{\xi_o}{I_o} = \frac{\xi_o / \sqrt{2}}{I_o \sqrt{2}} = \frac{\xi_{eff}}{I_{eff}} \qquad ...(1.33)$$

Thus, the inductive reactance may be defined as the ratio of root mean square (effective or virtue) emf across the inductance to the root mean square current flowing through it. The effect of inductive reactance $X_L = \omega L$ in an A C. circuit is the same as the resistance R in D.C. circuit to reduce the current in it, *but* without loss of *power*. The current I in the inductance L is obtained by dividing voltage across it by $j\omega L$. The impedance of the inductance is given by Eq. (1.30).

Equation (1.31) *shows that the current lags behind the voltage oy $\pi/2$* (or quarter of cycle) as shown in Fig. 1.8 (b). Fig. 1.8 (c) shows the complex number representation.

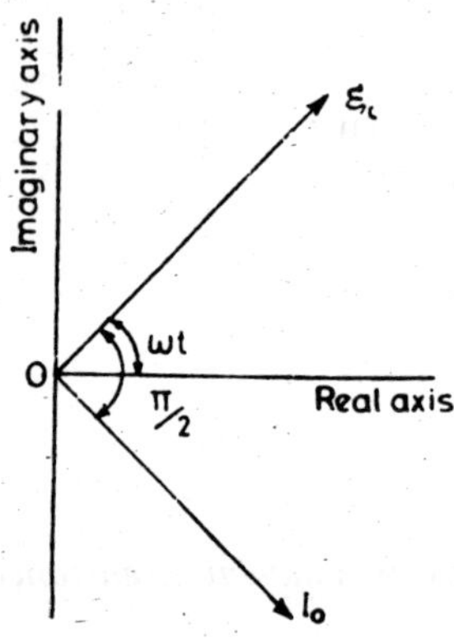

Fig. 1.8(c)

A.C. CIRCUIT CONTAINING CAPACITOR ONLY

Fig. 1.9 (a) shows a circuit containing a capacitor of capacitance C with an alternating voltage $\xi = \xi_0 \sin \omega t$ applied across it. This voltage is the imaginary part of the complex number $\xi_0 e^{j\omega t}$. Thus, in complex number representation, we can write,

$$\xi = \xi_0 e^{j\omega t}$$

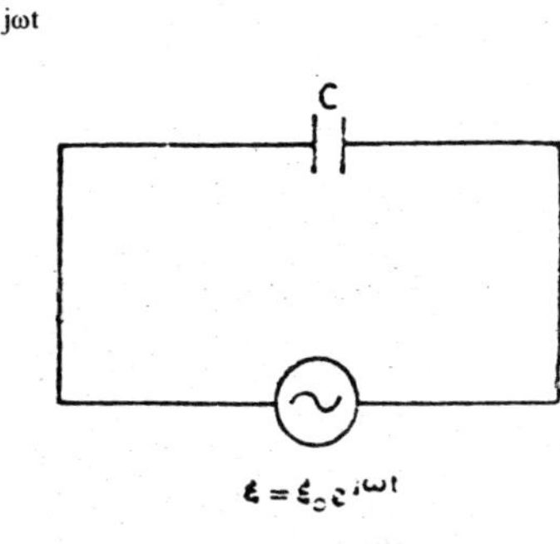

Fig. 1.9(a)

The instantaneous charge on the capacitor at any time t is,

$$q = C\xi = C\xi_0 e^{j\omega t} \qquad \text{...(1.34)}$$

Corresponding to this charge q, the instantaneous current which is in fact the displacement current due to the variation of voltage across the capacitor plates in the circuit is given by,

$$I = \frac{dq}{dt} = C\xi_0 \cdot j\omega e^{j\omega t} = \frac{\xi_0}{1/\omega C} j e^{j\omega t}$$

$$= \frac{\xi_0}{1/j\omega C} e^{j\omega t} \qquad \text{...(1.35)}$$

But $e^{j\pi/2} = \cos \pi/2 + j \sin \pi/2 = j$

Hence,

$$I = \frac{\xi_0}{1/\omega C} e^{j(\omega t + \pi/2)}$$

or, $$I = \frac{\xi_0}{X_C} e^{(\omega t \_ \pi/2)} = I_0 e^{j(\omega t + \pi/2)} \qquad \text{...(1.36)}$$

where $X_C = 1/\omega C$ is known as *capacitative reactance* and $I_0 = \xi_0/X_C$ is the *peak value of the current.*

Clearly equation (1.36) shows that the *current leads the voltage by π/2 as shown in Fig. 1.9 (b).*

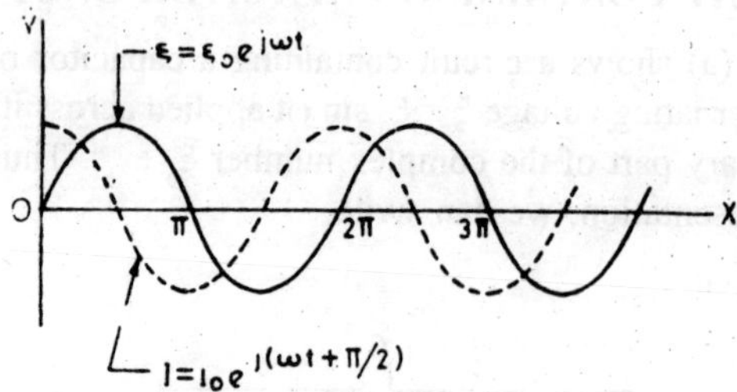

Fig. 1.9(b)

Further, the complex number representation is shown in Fig. 1.9 (a). The current I in the circuit is obtained by dividing exf applied by impedance of the capacitor (1/jωC) as shown in eq. (1.35).

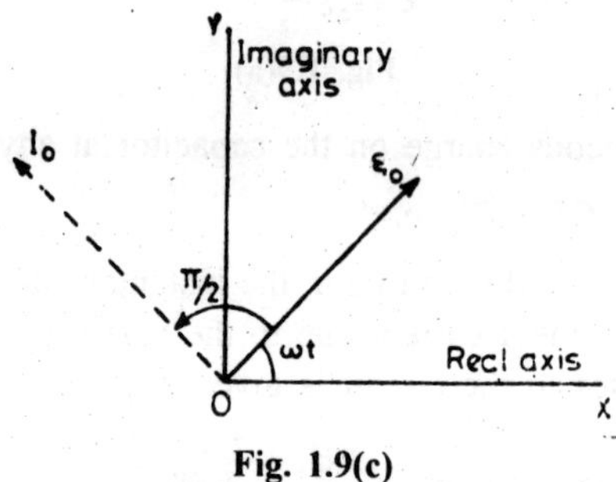

Fig. 1.9(c)

Since the capacitative reactance $X_C = 1/\omega C$ is inversely proportional to the frequency of applied voltage. Hence a condenser *offers negligible reactance at high frequency and works as "by-pass capacitor" and very high reactance at low frequency and works as "blocking condenser.*

A.C. CIRCUIT CONTAINING INDUCTANCE AND RESISTANCE IN SERIES

Let us consider a circuit as shown in Fig. 1.10 (a) containing a series combination of an inductance L and resistance R with an alternating emf x = xo ejwt applied across the combination. The instaneous current in the circuit is given by,

$$I = \frac{\text{Applied emf}}{\text{Vector impedance of the circuit}} \quad ...(1.37)$$

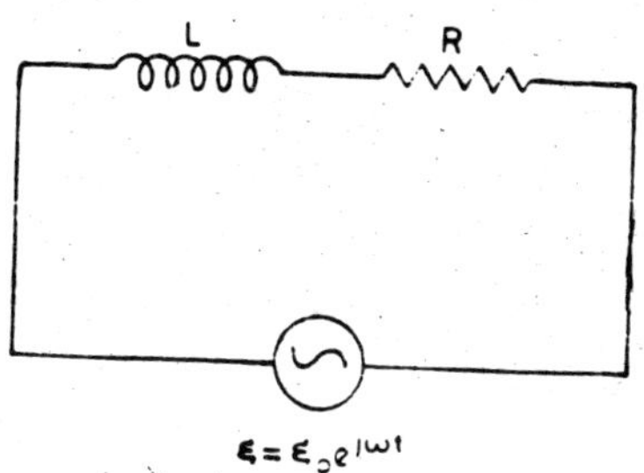

Fig. 1.10(a)

In the present case, Vector impedance $Z = R + j\omega L$ where R is impedance due to resistance and $j\omega L$ is the impedance due to inductance L. Z is also known as impedance operator. However,

Impedance $|Z|$ is given by,

$$|Z| = \sqrt{R^2 + \omega^2 L^2}$$

Thus the current,

$$I = \frac{\xi_0 e^{j\omega t}}{R + j\omega L} \quad \text{...(1.38)}$$

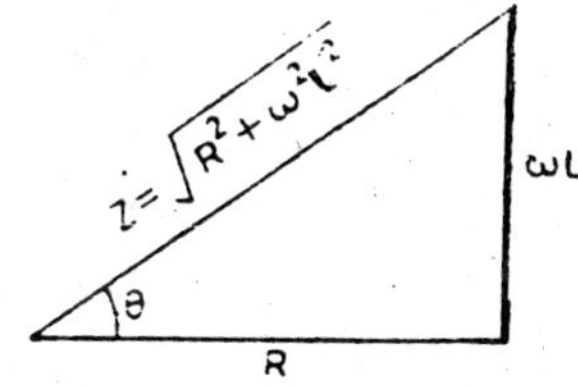

Fig. 1.10(b)

We can express $\frac{1}{R + j\omega L}$ as follows:

$$\frac{I}{R + j\omega L} = \frac{R - j\omega L}{(R + j\omega L)(R - j\omega L)}$$

$$= \frac{R - j\omega L}{R + \omega^2 L^2}$$

$$= \frac{R}{R^2 + \omega^2 L^2} - \frac{j\omega L}{R^2 + \omega^2 L^2} \quad \text{...(1.39)}$$

Let, $\propto = \frac{R}{R^2 + \omega^2 L^2}$

and $\beta = \frac{\omega L}{R^2 + \omega^2 L^2}$

Then, $\dfrac{1}{R + j\omega L} = \propto - j\beta$

Further put $\tan\theta = \dfrac{\omega L}{R}$, as shown in Fig. 1.10 (b).

Then $\cos\theta = \dfrac{R}{\sqrt{R^2 + \omega^2 L^2}}$ and

$$\sin\theta = \frac{\omega L}{\sqrt{R^2 + \omega^2 L^2}}$$

Then, $\dfrac{I}{R + j\omega L} = \propto - j\beta = \dfrac{1}{\sqrt{R^2 + \omega^2 L^2}}[\cos\theta - j\sin\theta]$

Hence eq. 1.38 becomes, $I = \dfrac{\xi_0 e^{j(\omega t - \theta)}}{\sqrt{R^2 + \omega^2 L^2}} = I_o\, e^{f(\omega t - \theta)}$ ...(1.40)

where impedance $Z = \sqrt{R^2 + \omega^2 L^2}$

and $I_o = \dfrac{\xi_o}{\sqrt{R^2 + \omega^2 L^2}}$

Equation (1.40) represents the variation of current in the circuit with time as shown in Fig. 1.10 (c). Equation (1.40) also indicates that the *current lags behind the applied emf in phase by an angle* θ where $\theta = \tan^{-1} \omega L/R$ as shown in Fig. 1.10 (d). In other Fig. 1.10(b)

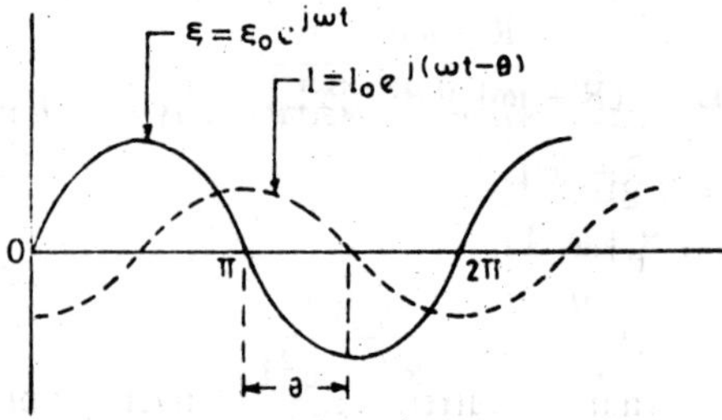

Fig. 1.10(c)

(or voltage) leads the current by angle θ. Fig. 1.10 (d) represents the complex number representation. However, the voltage ξ_R across the resistance is in phase with the current while voltage ξ_L across the inductance leading the current I by an angle π/2 and the total voltage,

$\xi_o = I_o \sqrt{R^2 + \omega^2 L^2}$ leading the current by an angle $\theta = \tan^{-1} \omega L/R$. The impedance

$$Z = \frac{\xi_o}{I_o} = \frac{\xi_o / \sqrt{2}}{I_o / \sqrt{2}} = \frac{\xi_{eff}}{I_{eff}}$$

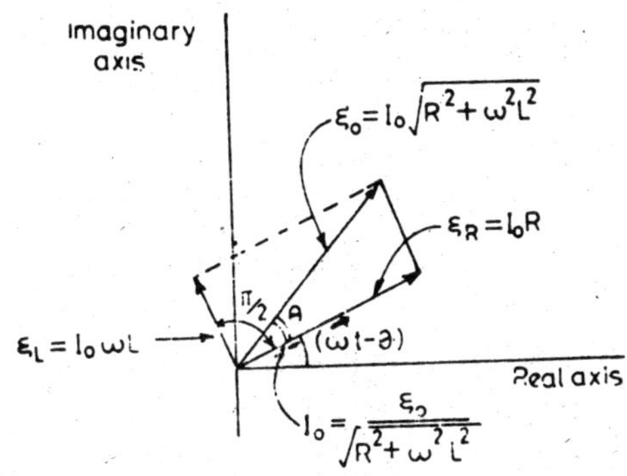

Fig. 1.10(d)

Thus, impedance may also be defined as the ratio of the r.m.s. (or virtual) value of the emf applied across the L – R circuit to the r.m.s. (or virtual) value of current flowing through the circuit.

Special Cases

1. If $R \to 0$ *i.e.,* when inductive reactance ωL is very large than to ohmic resistance R which can be neglected. Then $Z = \omega L$ and $\theta = \pi/2$.

   In *practice, it occurs when the frequency is very high.*

   Then current is given by,

$$I = I_o\, e^{j(\omega t - \pi/2)}$$

and $\qquad I_o = \xi_o/\omega L$

Thus the peak value of current *i.e.,* I_o is inversely proportional to both self inductance and frequency. Therefore, when the frequency is high, a high voltage can produce only small current.

Further since phase lag $\theta = \tan^{-1} \dfrac{\omega L}{R}$.

Hence, greater the reactance compared with the ohmic resistance the phase lag increases. Thus, the current and the emf can be brought into phase lag increases. Thus, the current and the emf can be brought into phase by introduction of a non-inductive resistance. Practically this is adopted in A.C. voltmeters and wattmeters.

2. If $\omega L \rightarrow 0$ *i.e.*, when inductance or the frequency ($f = \omega/2\pi$) is so small as to be negligible, then ωL will be very small compared to R and we get,

$$\theta \approx 0 \text{ and } Z = R$$

$$\text{and } I = I_o\, e^{j\omega t} = \frac{\xi_o e^{j\omega t}}{R} = \frac{\xi}{R}$$

A.C. CIRCUIT CONTAINING A RESISTANCE AND A CAPACITANCE IN SERIES

Fig. 1.11 (a) shows a circuit consisting of a series combination of a resistance R and a capacitor C with an alternating emf $\xi = \xi_o\, j\omega t$ applied across the combination. The instantaneous current I is given by,

$$I = \frac{\text{Applied emf}}{\text{Vector impedance of the circuit}}$$

$$= \frac{\xi_o e^{j\omega t}}{R + \dfrac{1}{j\omega C}} = \frac{\xi_o e^{j\omega t}}{R - \dfrac{j}{\omega C}} \qquad \text{...(1.41)}$$

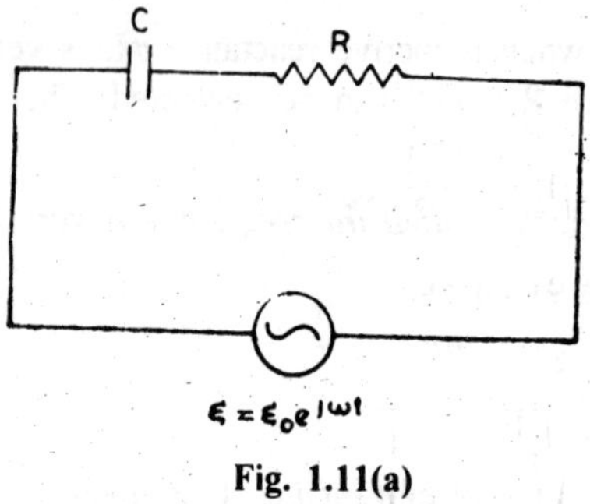

Fig. 1.11(a)

where R is impedance due to resistance and $1/j\omega C$ is the impedance due to capacitance C. As in the previous case, we can express,

$$\frac{I}{R - \dfrac{j}{\omega C}} = \frac{R + \dfrac{j}{\omega C}}{R^2 + \dfrac{1}{\omega^2 C^2}}$$

$$= \frac{R}{R^2 + \frac{1}{\omega^2 C^2}} + \frac{j/\omega C}{R^2 + \frac{1}{\omega^2 C^2}}$$

Put $\qquad \propto = \dfrac{R}{R^2 + \frac{1}{\omega^2 C^2}}$

and $\qquad \beta = \dfrac{1/\omega C}{R^2 + \frac{1}{\omega^2 C^2}}$

and $\qquad \tan\theta = \dfrac{1}{\omega CR}$

We get, $\dfrac{I}{R - \frac{j}{\omega C}} = \alpha + j\beta$

$$= \frac{1}{\sqrt{R^2 + \frac{1}{\omega^2 C^2}}}[\cos\theta + j\sin\theta]$$

$$= \frac{1}{\sqrt{R^2 + \frac{1}{\omega^2 C^2}}} e^{j\theta}$$

Hence Eq. (1.41) reduces to

$$I = \frac{\xi_o e^{j(\omega+\theta)}}{\sqrt{R^2 + \frac{1}{\omega^2 C^2}}}$$

or $\qquad I = I\, e^{j(\omega + \theta)}$ ...(1.42)

where, $\quad I_o = \dfrac{\xi_o}{\sqrt{R^2 + \frac{1}{\omega^2 C^2}}}$

indicates the *peak value of current.* Equation (1.42) represents the variation of current with time and shows that *current leads the applied voltage in phase by an angle* θ as shown in Fig. 1.11 (b) given, by,

$$\theta = \tan^{-1} \frac{1/\omega C}{R}$$

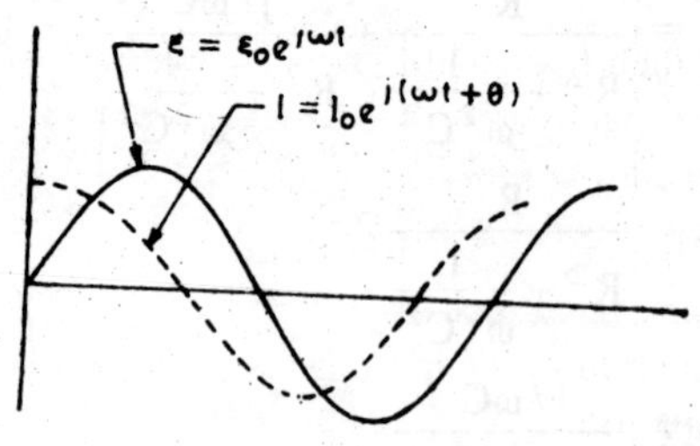

Fig. 1.11(b)

$$\text{The impedance, } Z = \sqrt{R^2 + \frac{1}{\omega^2 C^2}}$$

$$= \sqrt{(\text{Resistance})^2 + (\text{Capacitative reactance})^2}$$

Fig. 1.11 (c) shows the complex number representation of current and voltage. *The current leads the applied voltage by phase is the same as the voltage lags behind the current in phase.* The voltage ξ_R across the resistance is in phase with current, while the voltage ξ_C across the condenser lagging the current I by an angle $\pi/2$. Finally the total voltage,

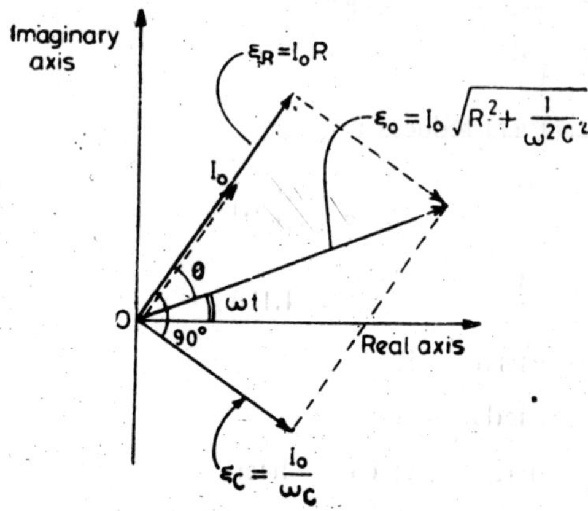

Fig. 1.11(c)

$$\xi_0 = I_0 \sqrt{R^2 + \frac{1}{\omega^2 C^2}}$$

lagging the current I by an angle $\theta = \tan^{-1} \dfrac{1/\omega C}{R}$.

Special Cases

1. If C → 0, then I_0 → 0, this means no current will flow in the circuit with zero capacity and this causes a break in the circuit. The same result will be obtained with very low frequencies.
2. If C ∞ ∞, then $I_0 = \xi_0/R$. Thus a condenser of infinite capacity does not offer any resistance to A.C. and tan θ = 0 or θ → 0. This means the introduction of such a condenser causes no phase change. The same result follows if the frequency is very high.
3. If R → 0, then $I = \xi_0 \omega C$ and tan q → ∞ or θ = π/2. Thus the current is leading by π/2 in phase with respect to emf.

A.C. CIRCUIT WITH RESISTANCE, CAPACITANCE AND INDUCTANCE IN SERIES

Let a sinusoidal alternating emf $\xi = \xi_0\, e^{j\omega t}$ be applied to the series combination of L.C.R. as shown in Fig. 1.12 (a). The vector impedance of the combination is sum of three parts.

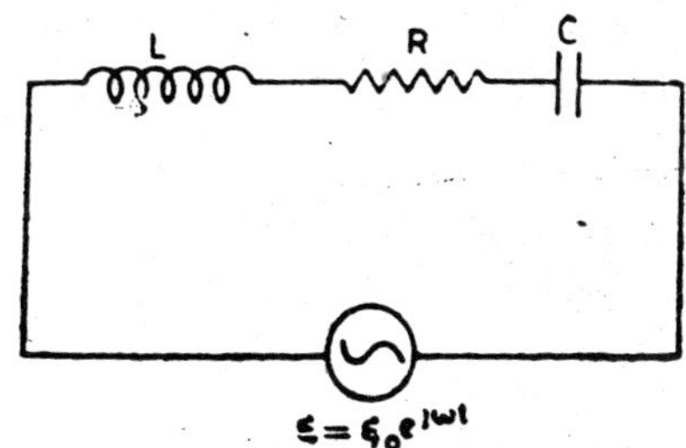

Fig. 1.12(a)

1. R due to resistance R
2. jωL due to inductance L
3. 1/jωC = – jωC due to capacitance C.

Thus in vector form the emf eqn. of the circuit is

$$\left[R + j\left(\omega L - \frac{1}{\omega C}\right)\right] I = \xi_0\, e^{j\omega t} \qquad ...(1.43)$$

where I represents the instantaneous current in circuit, is given by,

$$I = \frac{\xi_0 e^{j\omega t}}{R + j\left(\omega L - \frac{1}{\omega C}\right)} \qquad ...(1.44)$$

Adopting the same procedure as in the previous article, we can write,

$$\frac{1}{R + j\left(\omega L \frac{1}{\omega C}\right)} = \frac{R - j\left(\omega L - \frac{1}{\omega C}\right)}{R^2 + \left(\omega L - \frac{1}{\omega C}\right)^2}$$

$$= \propto - j\beta$$

where, $x = \dfrac{R}{R^2 + \left(\omega L - \frac{1}{\omega C}\right)^2}$

and $\beta = \dfrac{\omega L - \frac{1}{\omega C}}{R^2 + \left[\omega L - \frac{1}{\omega C}\right]^2}$

and if we put,

$$\tan \theta = \frac{\omega L - \frac{1}{\omega C}}{R}$$

Then, $\cos \theta = \dfrac{R}{\sqrt{R^2 + \left(\omega L - \frac{1}{\omega C}\right)^2}}$

and $\sin \theta = \dfrac{\omega L - 1/\omega C}{\sqrt{R^2 + \left(\omega L - \frac{1}{\omega C}\right)^2}}$

Hence,

$$\frac{1}{R + j\left(\omega L - \frac{1}{\omega C}\right)} = \frac{1}{\sqrt{R^2 + \left(\omega L - \frac{1}{\omega C}\right)^2}}(\cos\theta - j\sin\theta)$$

$$= \frac{1}{\sqrt{R^2 + \left(\omega L - \frac{1}{\omega C}\right)}} e^{-j\theta}$$

Then the expression for current *i.e.,* eq. (1.44) reduces,

$$I = \frac{\xi_o}{\sqrt{R^2 + \left(\omega L - \frac{1}{\varepsilon C}\right)^2}} e^{j(\omega t - \theta)}$$

or $\quad I = I_o \, e^{j(\omega t - \theta)}$ ...(1.45)

where, $I_o = \dfrac{\xi_o}{\sqrt{R^2 + \left(\omega L - \frac{1}{\omega C}\right)^2}}$ ...(1.46)

I_o is the *peak value of the current.* Ans.

$$\sqrt{R^2 + \left(\omega L - \frac{1}{\omega C}\right)^2} = Z$$

where Z is called the *impedance* of the circuit (in magnitude), which is the ratio of peak emf to the peak current. Again impedance, from eq. (1.46), is given by

$$Z = \frac{\xi_o}{I_o} = \frac{\xi_o / \sqrt{2}}{I_o \sqrt{2}} = \frac{\xi_{eff}}{\xi_{eff}}$$

Thus, the impedance of a L.C.R. circuit can be defined as the ratio of the r.m.s. value of emf applied to the r.m.s. current flowing through it.

Eq. (1.45) shows that the *current lags the applied voltage in phase by an angle* θ *given by,*

$$\theta = \tan^{-1} \frac{\omega L - \frac{1}{\omega C}}{R}$$

The variation of current and voltage with time is shown in Fig. 1.12(b). In other words, the emf and current have a phase difference of θ.

This phase difference is shown in Fig. 1.12 (c) in the complex plane. The voltage across the resistance ξ_R is shown in phase with the current while,

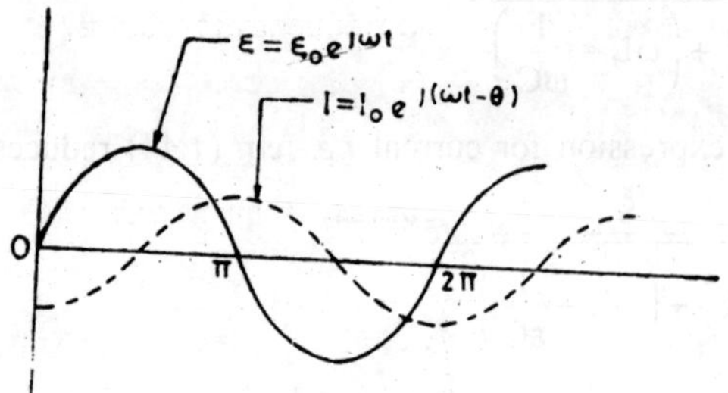

Fig. 1.12(b)

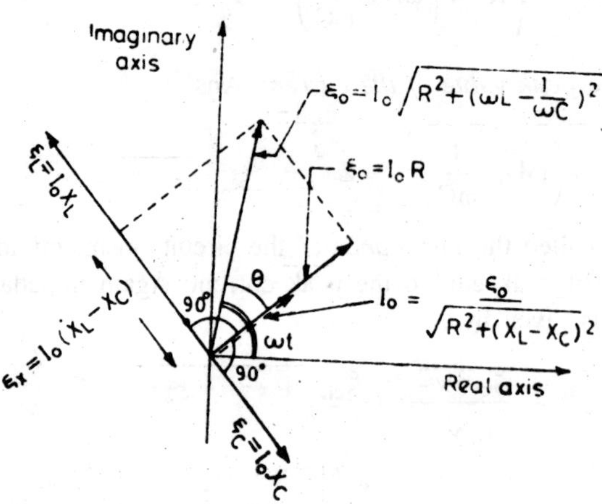

Fig. 1.12(c)

which is the net voltage across the combination of L and C and leads the current by an angle $\pi/2$. The total voltage ξ_o is shown leading the current by an angle,

$$\theta = \tan^{-1} \frac{X_L - X_C}{R}$$

Special Cases

The phase difference (*i.e.*, leading or lagging) in fact depends on the value of θ which depends on value of X_L and X_C. Depending on the values of X_L and X_C there are three important cases.

1. If $\omega L > \frac{1}{\omega L}$ *i.e.,* when impressed frequency ($f = \omega/2\pi$) is very large, the phase angle θ will be positive and the current will lag behind the applied emf. The potential difference $I_0 X_L = I_0 \omega_L$ across the inductance is greater than potential difference $I_0/\omega C$ across the capacitance and the circuit behaves as an *inductive circuit.* The vector voltage in complex plane is shown in Fig. 1.12 (c).

2. If $\omega L < \frac{1}{\omega C}$, the phase angle θ will be negative and the current will lead the applied emf. The circuit behaves as *capacitative circuit.* The vector diagram for the case is shown in Fig. 1.12 (d) in complex plane.

3. If $\omega L = 1/\omega C$ *a very important and interesting case arises.* In this case,

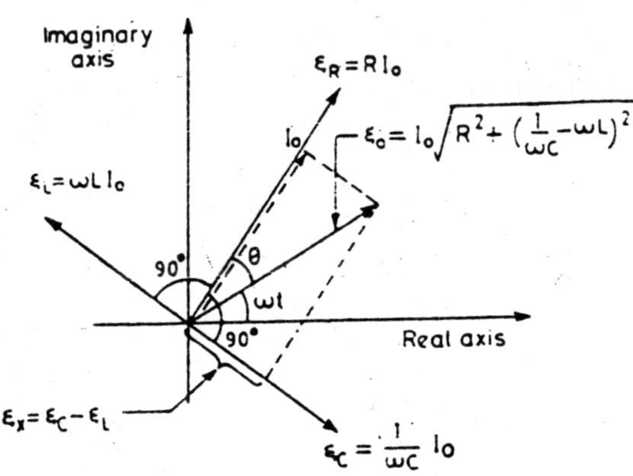

Fig. 1.12(d)

$$\theta = \tan^{-1} \frac{\omega L - \frac{1}{\omega C}}{R} = 0$$

i.e., the phase angle θ becomes zero and the emf and current will be in phase. *The potential difference across the inductance and capacitance are equal in magnitude but opposite in phase and therefore cancel out* and the whole voltage is dropped across the resistance. Therefore, the circuit behaves as *pulley resistive circuit.* The vector diagram in complex plane is shown in Fig. 1.12 (e).

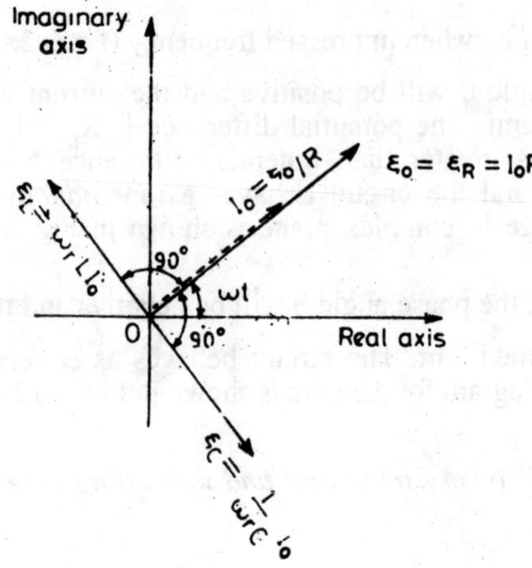

Fig. 1.12(e)

Under this condition, the peak current will be maximum given by I_o max

$$I_{0max} = \frac{\xi_o}{\sqrt{R^2 + \left(\omega L - \frac{1}{\omega C}\right)^2}} = \frac{\xi_o}{R} \qquad ...(1.47)$$

Thus, in this particular case when $\omega L = 1/\omega C$, the emf and current will be in phase and *current is maximum*. The circuit is said to be a *series resonant circuit* and the phenomenon of maximum current is called *resonance*. Thus at resonance,

$$\omega_r L = 1/\omega_r C$$

or $$\omega_r = 1/\sqrt{LC}$$

or $$f_r = \frac{1}{2\pi\sqrt{LC}} \qquad ...(1.48)$$

Here subscript r corresponds to resonance.

The frequency at which resonance occurs is known as *resonance frequency*. Evidently from Eq (1.48) the resonant frequency is the same as the frequency of the oscillatory discharge of a series LCR circuit when the resistance is low. Thus, the series LCR circuit is in resonance with applied voltage. The *frequency of applied voltage coincides with the natural frequency of the circuit*. Fig. 1.12 (e) gives the vector diagram of voltage and current for this circuit on the complex plane under this condition Fig. 1.12 (f) shows the variation of peak current with frequency for resonant circuit.

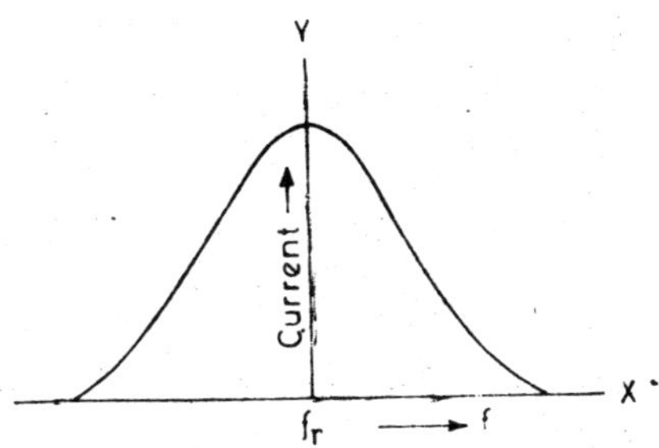

Fig. 1.12(f)

SHARPNESS OF RESONANCE OF A SERIES LCR CIRCUIT

We have discussed in the previous article that the current amplitude is,

$$I_o = \frac{\xi_o}{\sqrt{R^2 + \left(\omega L - \frac{1}{\omega C}\right)^2}}$$

At resonance $\omega L = 1/\omega C$

therefore $I_{omax} = \xi_o/R$

Let us examine the effect of circuit resistance on the current when the supply frequency is varied continuously from a value below the resonant frequency to above it, when the resistance of the circuit is (1) low, (2) medium and (3) high. The curves obtained by plotting amplitude of current (or voltage drop across a fixed resistance) against ω (or f) are known as *resonance curves.*

Fig. 1.13 (a) shows the variation of current amplitude as function of frequency for three different values of resistance R.

Clearly, as the resistance in the circuit is reduced the resonance curve becomes sharper. The peak value of I_o shown that the *circuit responds only to the frequency exactly equal to its natural frequency of the circuit* ω_r ($\omega_r = 1/\sqrt{LC}$ for low resistance in LCR circuit) and to none others. The resonance here is, therefore, said to be *sharp.*

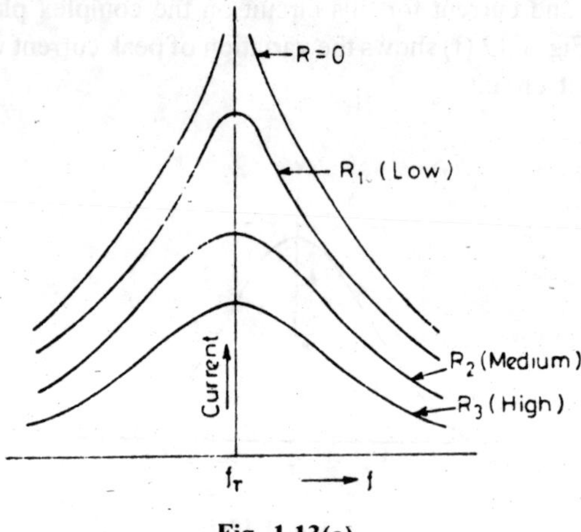

Fig. 1.13(a)

Thus, *sharpness of resonance is, in a way, a measure of the* rate of *fall of amplitude from its maximum value at resonant frequency on either side of it.* The more quickly the current amplitude falls for changes of frequency f_r the sharper is said to be resonance.

On the other hand, if the amplitude remains more or less at it peak value over an appreciable range of frequency on either side o $\omega_r = \frac{1}{\sqrt{LC}}$ resonant frequency *i.e.*, even when $\omega_r \neq \frac{1}{\sqrt{LC}}$. Thus, the *circuit responds to a number of frequencies* near about f_r on either side of it. The resonance in this case, is therefore, said to be *flat.*

Variation of Voltages Across Inductance and Capacitance is a Series L.C.R. Resonant Circuit

The voltage drop across the inductance coil is

$$\xi_L = I_o X_L = \frac{\xi_o}{Z} X_L = \frac{\xi_o}{\sqrt{R^2 + \omega^2 L^2}} \omega L$$

For frequency considerably above resonant frequency $\omega L > R$, we have,

$$\xi_L = \frac{\xi_o \omega L}{\omega L} = \xi_o$$

equal to *amplitude of impressed voltage.*

Similarly, the voltage drop across the capacitor is,

$$\xi_C = I_o X_C = \frac{\xi_o}{\sqrt{R^2 + \frac{1}{\omega^2 C^2}}} \cdot \frac{1}{\omega C}$$

For very low frequencies approaching zero, R^2 can be neglected in comparison with $1/\omega^2 C^2$, we have,

$$\xi_C = \frac{\xi_o}{1/\omega C} \cdot \frac{1}{\omega C} = \xi_o$$

equal to the impressed voltage.

Fig. 1.13 (b) shows the variation of ξ_L and ξ_C with ω. Both of these curves are similar in shape in the vicinity of the resonance to the curves of Fig. 1.13 (a). This is because the resonance phenomenon takes place in a very limited range. The reactance across which the voltage is developed remains substantially the same in the region round about the resonance. The product of current and reactance, therefore, varies with frequency practically in the same way as current.

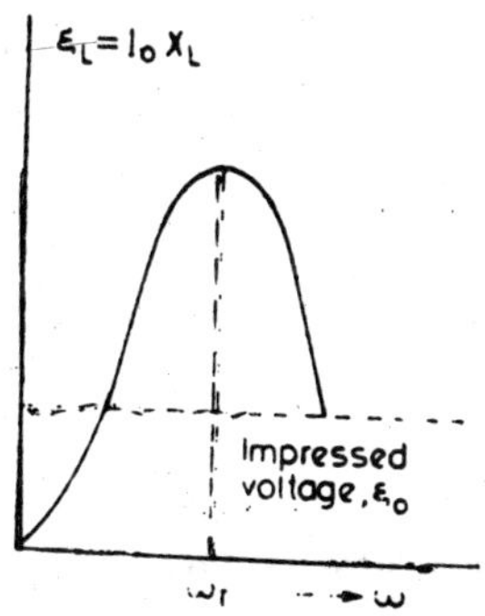

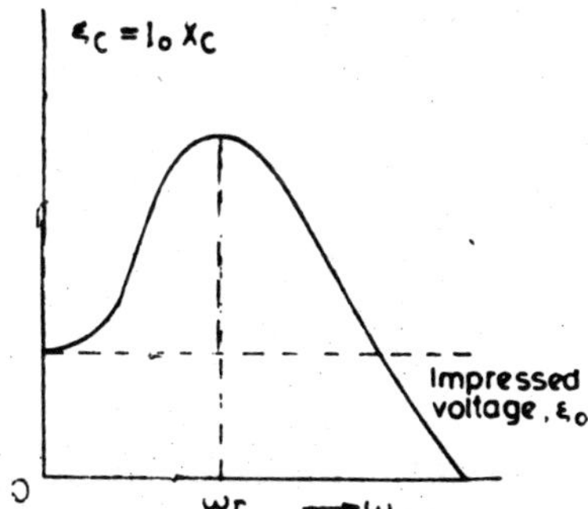

Fig. 1.13(b)

Q of a Circuit

Qualitatively the sharpness of resonance curve is determined by a quality factor called "Q" of the circuit. It is *defined as there ratio of reactance of either the inductance or capacitance at the resonant frequency to the total resistance of the circuit.* Thus, mathematically,

$$Q = \frac{X_L}{R} = \frac{\omega_r L}{R}$$

or $$Q = \frac{X_C}{R} = \frac{1}{\omega_r CR} \quad ...(1.49)$$

Since at resonance $X_L = X_C$, *i.e.*, $\omega_r L = 1/\omega_r C$, we get the same value of Q from both expressions.

How the sharpness depends on "Q" can be understood as follows. Since at resonance the current is inversely proportional to the resistance ($I_o = \xi_o/R$). So, directly proportional to Q (see Eq. 1.49). The value of the current for frequencies appreciably different form resonant frequency is smainly determined by the reactance and not by the resistance. Thus the current is nearly independent of the resistance of the circuit. The result of increasing the resistance of the circuit (*i.e.*, lowering Q) reduces the reasons at the resonant frequency but leaves the other frequencies nearly unaffected. Therefore, if Q is large the resonance curve is sharp.

"Q" is also defined in terms of *lower and upper half-power frequencies.* The half-power frequencies are the frequencies at which the power dissipation in the circuit drops to one half its resonance value. If f_1 and f_2 are the lower and upper half power frequencies and f_r is the resonant frequency,

The power at resonant frequency $= I^2_{omax}$

$$= \left(\frac{\xi_o}{R}\right)^2 R_o$$

The power dissipation at f_1 is given by,

$$I_1^2 R = \frac{1}{2}\left(\frac{\xi_o}{R}\right)^2 R$$

where I_1 is current at lower half power frequency

or $$I_1 = \frac{1}{\sqrt{2}} \cdot \frac{\xi_o}{R} = \frac{1}{\sqrt{2}} \text{ current at resonance.}$$

Similarly, I_2 the current at upper half power frequency is

$$I_2 = \frac{1}{\sqrt{2}} \frac{\xi_o}{R} = \frac{1}{\sqrt{2}} \text{ current at resonance}$$

Thus at lower or upper half power frequencies the current amplitude becomes $1/\sqrt{2}$ times the current amplitude at resonance.

Hence from eq. (1.47) at lower or upper half power frequency, we have,

$$\frac{\xi_0}{\sqrt{R^2+\left(\omega L-\frac{1}{\omega L}\right)^2}} = \frac{1}{\sqrt{2}}\cdot\frac{\xi_0}{R}$$

On squaring and rearranging, we get,

$$R^2+\left(\omega L-\frac{1}{\omega C}\right)^2 = 2R^2$$

or $$\left(\omega L-\frac{1}{\omega L}\right)^2 = R^2$$

This equation has two solutions one for lower half and other for upper half power frequencies. Thus,

$$\omega_1 L - \frac{1}{\omega_1 C} = -R \qquad ...(1.50)$$

$$\omega_2 L - \frac{1}{\omega_2 C} = +R \qquad ...(1.51)$$

Adding these two equations, we get,

$$(\omega_1+\omega_2)\,L - \frac{1}{C}\left(\frac{1}{\omega_1}+\frac{1}{\omega_2}\right) = 0$$

or $$(\omega_1+\omega_2)\,L - \frac{\omega_1+\omega_2}{\omega_1\omega_2}\frac{1}{C} = 0$$

or $$\omega_1\,\omega_2 = \frac{1}{LC} \qquad ...(1.52)$$

or $$\omega_1\,\omega_2 = \omega_r^2 \text{ (by eq. 1.48)} \qquad ...(1.53)$$

Thus, the resonant frequency f_r is the geometric mean of the upper and lower half power frequencies.

Subtracting Eq. (1.50) from Eq. (1.51), we get,

$$(\omega_2-\omega_1)\,L + \frac{1}{C}\left(\frac{1}{\omega_1}-\frac{1}{\omega_2}\right) = 2R$$

$$(\omega_1-\omega_2)\,L + \frac{\omega_2-\omega_1}{\omega_1\omega_2}\frac{1}{C} = 2R$$

Substituting $\omega_1\,\omega_2$ from Eq. (1.52),

$$(\omega_2 - \omega_1)\left(L + \frac{1}{1/LC}\frac{1}{C}\right) = 2R$$

$$\omega_2 - \omega_1 = R/L$$

Therefore,

$$\frac{\omega_r}{\omega_2 - \omega_1} = \frac{f_r}{f_2 - f_1} = \frac{\omega_r L}{R} = Q \qquad ...(1.54)$$

Q in terms of *band width* is expressed as,

$$Q = \frac{\omega_r}{2\nabla\omega} \qquad ...(1.55)$$

where $\nabla\omega$ is the change in ω on either side of ω_r as a result of which the power dissipation becomes half its value at the resonant frequency. Thus,

$$\omega_1 = \omega_2 - \nabla\omega$$

and $$\omega_2 = \omega_r + \nabla\omega$$

The frequency difference,

$$\omega_2 - \omega_1 = 2\nabla\omega$$

$2\nabla\omega$ is called the *band width.*

Thus, greater value of Q corresponds to smaller $\omega_2 - \omega_1$ or $2\nabla\omega$ band width, the sharper is the resonance.

Selectivity

If the applied alternating voltage has a number of frequency components (for example, the signal received by a radio antenna has wide range of frequencies). Then a series L.C.R. circuit will give a maximum response (*i.e.,* it will pass a maximum current and have a maximum potential difference across its inductance) for only that frequency component which has the frequency,

$$f_r = \frac{1}{2\pi\sqrt{LC}}$$

Thus out of number of available frequencies it selects one frequency for which the current is maximum and for other frequencies the current is comparatively very small, in other words it shows selectivity. The resonant circuit is consequently known as the "acceptor circuit" as it practically accepts one frequency component of the applied voltage and rejects other. It may be noted that sharper the response (*i.e.*, higher the Q value) the more highly selective the acceptor circuit is.

Voltage Magnification

The circuit Q is also a measure of voltage magnification in the series L.C.R. circuit which can be understood as follows.

At resonance the potential difference across inductance and capacitance are equal and 180° out of phase and hence cancel. Therefore, the only potential difference at resonance is across the resistance.

As resonant current is maximum and given by,

$$I_{omax} = \frac{\xi_o}{R}$$

potential difference across resistance $R = I_{omax} R = \xi_o$

Thus, at resonance the potential difference across resistance is equal to the applied emf. The voltage magnification in L.C.R. circuit is defined as,

$$\text{Voltage magnification} = \frac{\text{p.d. across the induc tan e or capaci tan e}}{\text{Spplied voltage}}$$

$$m = \frac{I_{o\,max}\,\omega_r L}{\xi_o}$$

$$= \frac{\xi_o}{R}\frac{\omega_r L}{\xi_o} = \frac{\omega_r L}{R} = Q = \frac{1}{R}\sqrt{\frac{L}{C}} \qquad \text{...(1.56a)}$$

Similarly for capacitance,

$$m = \frac{I_{o\,max}}{\xi_o}\frac{1}{\omega_r C}$$

$$= \frac{\xi_o}{R\xi_o}\frac{1}{\omega_r C}$$

$$= \frac{1}{\omega_r CR} = Q = \frac{1}{R}\sqrt{\frac{L}{C}} \qquad \text{...(1.56b)}$$

Thus, the voltage magnification of L.C.R. circuit is equal to its quality factor Q at resonance.

A PARALLEL (OR ANTI) RESONANT CIRCUIT

Let an alternating emf $\xi = \xi_o e^{j\omega t}$ be applied across a coil of inductance L of negligible resistance in parallel with a capacitor of capacitance C as shown in Fig. 1.14 (a). The current through the inductance lags behind the applied emf in phase by $\pi/2$ and is given by equation (1.30),

$$I_L = \frac{\xi_o}{j\omega L} e^{j\omega t} = \frac{\xi_o}{\omega L} e^{j(\omega t - \pi/2)} \text{ as } \frac{1}{j} = -j = e^{-j\pi/2}$$

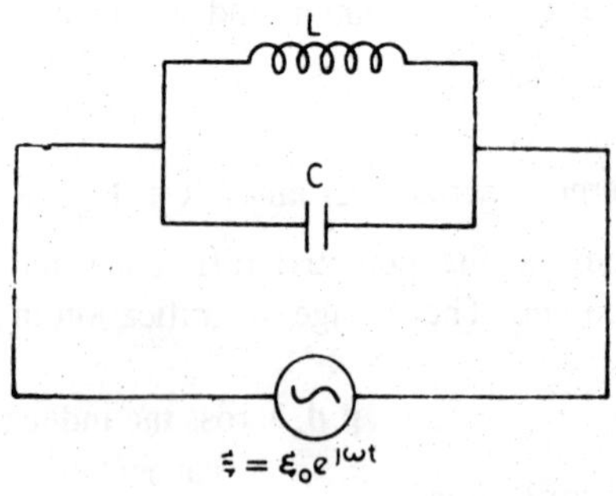

Fig. 1.14(a)

The current through the capacitor leads the applied emf in phase by $\pi/2$ and is given by Eq. (1.35),

$$I_C = \frac{\xi_o}{\frac{1}{j\omega C}} e^{j\omega t} = \frac{\xi_o}{\frac{1}{\omega C}} e^{j(\omega t + \pi/2)}$$

$(j = e^{j\pi/2})$

therefore the current in the two branches differ in phase by π. The total current I_T will be the vector sum I_L and I_C and will be equal to their numerical difference, since they are 180° out of phase. Thus,

$$I_T = I_L + I_C$$

$$= \frac{\xi_o}{j\omega L} e^{j\omega t} + j\xi_{o\omega} C e^{j\omega t}$$

$$= j\left(\varepsilon C - \frac{1}{\omega L}\right)\xi_o e^{j\omega t} \qquad ...(1.57)$$

The phase relationship between applied voltage and current is shown in Fig. 1.14 (b) in complex plane.

The total impedance of the parallel circuit is given by,

$$\frac{1}{Z_T} = \frac{1}{Z_L} + \frac{1}{Z_C} = \frac{Z_L + Z_C}{Z_L Z_C}$$

or $$Z_T = \frac{Z_L Z_C}{Z_L + Z_C}$$

$$Z_T = \frac{j\omega L \dfrac{1}{j\omega C}}{j\left(\omega L - \dfrac{1}{\omega C}\right)}$$

$$= \frac{L/C}{j\left(\omega L - \dfrac{1}{\omega C}\right)}$$

$$= \frac{L/C}{\dfrac{jL}{C}\left(\omega C - \dfrac{1}{\omega L}\right)}$$

$$= \frac{1}{j\left(\omega C - \dfrac{1}{\omega L}\right)}$$

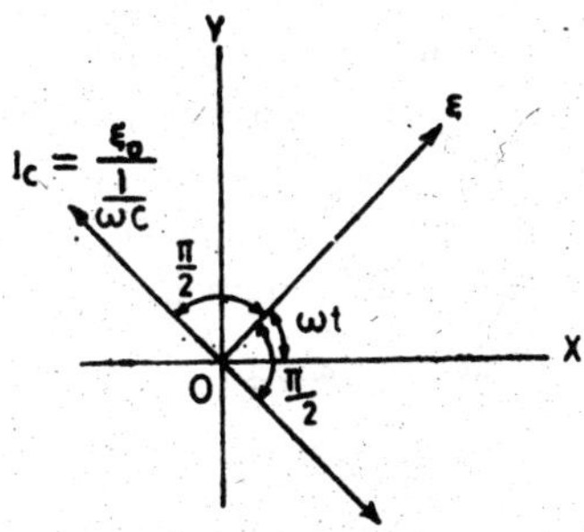

Fig. 1.14(b)

by equation (1.57),

$$Z_T = \frac{\xi_0 e^{j\omega t}}{I_T} \qquad ...(1.58)$$

The resonance is obtained when the current in the two branches is equal. Since current is out of phase in two branches, the total current will be zero. Hence by Eq. (1.57), at resonance

$$\omega_r C - \frac{1}{\omega_r L} = 0$$

or $$\omega_r^2 = \frac{1}{LC}$$

The resonant frequency $f_r = \dfrac{1}{2\pi\sqrt{LC}}$

This is identical with the series resonant case.

When I_T is zero. $Z_T = \infty$

Thus, at the frequency f_r of the supply voltage, the circuit offers infinite impedance to the flow of the current so that no current is drawn from the supply. Such a circuit offering infinite impedance to the flow of current is called *parallel resonant circuit* and the particular frequency f_r is called *resonant frequency.*

The *paradox of current flowing* in the coil and the condenser while no current is being supplied by the source is explained by the fact that circuit is now oscillating and the condenser is alternately charging and discharging through the coil in synchronised manner with the impressed frequency. As the coil and condenser have no resistance, the amount of energy taken by the condenser for initial charge from the source is not dissipated. The system is analogous to a frictionless pendulum which will oscillate continuously upon receiving as initial displacement without further energy being supplied.

Further at resonance,

$$Z_T = \frac{Z_L\ Z_C}{Z_L + Z_C}$$

$Z_L + Z_C$ is the total impedance when L and C are connected in series. Let us denote it by Z_S, hence we get,

$$Z_T = \frac{j\omega_r L \dfrac{1}{j\omega_r C}}{\dfrac{j\omega_r C}{Z_S}}$$

but at resonance $\omega_r L = 1/\omega_r C$

Therefore,

$$Z_T = \frac{(\omega_r L)^2}{Z_S}$$

multiplying numerator and denominator by $\xi_o\ e^{j\omega t}$, we get

$$Z_T = \frac{(\omega_r L)^2}{\xi_o e^{j\omega t}} \cdot \frac{\xi_o e^{j\omega t}}{Z_S}$$

For given supply voltage and circuit,

$\dfrac{\omega_r L}{\xi_o e^{j\omega t}}$ is constant. Hence,

$$Z_T = \text{const. } I_S \qquad \text{...(1.59)}$$

where $I_S = \dfrac{\xi_0 e^{j\omega t}}{Z_S}$ is the current when L and C are connected in series with the same emf applied. *Equation (1.59) shows that the impedance variation with frequency in parallel resonance case has exactly the same shape as the current variation with frequency in series resonance case.*

PARALLEL RESONANT CIRCUIT WHEN INDUCTANCE L HAVE SOME RESISTANCE

In actual practice some resistance is always associated with inductance used. Suppose an alternating emf ξ is applied to a capacitance C and inductance L having a resistance R connected in parallel as shown in Fig. 1.15 (a). *Because of resistance, the current through the inductive coil will lag behind the applied emf by an angle less than $\pi/2$ and so the phase difference between I_L and I_C will be less than 180°.* Due to this the resultant impedance of the combination will be maximum at resonance instead of being infinite because of the presence of resistance in the circuit. Clearly the problem of defining parallel resonance becomes defining parallel resonance becomes more complicated. There are two criterion for solving such problems.

1. The condition of maximum impedance and
2. The condition of unity power factor.

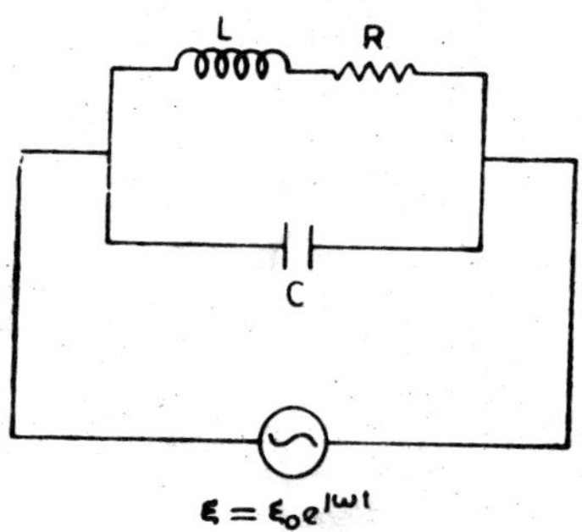

Fig. 1.15(a)

In order to avoid confusion parallel resonance is usually taken to mean the condition for maximum impedance is called anti-resonance. *Anti in the sense that in series circuit at resonance the impedance is minimum.*

Let the combination shown in Fig. 1.15 (a) be connected to A.C. emf given by,

$$\xi = \xi_o e^{j\omega t}$$

The total impedance Z_T for the combination is given by,

$$\frac{1}{Z_T} = \frac{1}{Z_L} + \frac{1}{Z_C} = \frac{1}{R + j\omega L} + j\omega C$$

$$\frac{1}{Z_T} = \frac{1 + j\omega C R - \omega^2 LC}{R + j\omega L}$$

or $$Z_T = \frac{R + j\omega L}{(1 - \omega^2 LC) + j\omega CR}$$

$$= \frac{R + j\omega L}{(1 - \omega^2 LC) + j\omega CR} \cdot \frac{(1 - \omega^2 LC) - j\omega CR}{(1 - \omega^2 LC) - j\omega CR}$$

$$= \frac{R + j\omega [L - C(\omega^2 L^2 + R^2)]}{(1 - \omega^2 LC)^2 + \omega^2 C^2 R^2}$$

$$= R_o + j\omega L_o \text{ (say)} \qquad ...(1.60)$$

where,

$$R_o = \frac{R}{(1 - \omega^2 LC)^2 + \omega^2 C^2 R^2} \qquad ...(1.61)$$

represents the effective or apparent resistance of the coil and,

$$L_o = \frac{L - C(\omega^2 L^2 + R^2)}{(1 - \omega^2 LC^2) + \omega^2 C^2 R^2} \qquad ...(1.62)$$

represents the effective inductance of the coil.

The magnitude of the impedance of the combination is given by,

$$Z = |ZT| = \sqrt{R_o^2 + \omega^2 L_o^2}$$

or $$Z^2 = \omega^2 [L - C(\omega^2 L^2 + R^2)]^2$$

After simplification

$$Z^2 [(1 - \omega^2 LC)^2 + \omega^2 C^2 R^2]^2$$

$$= R^2 [(1 - \omega^2 LC)^2 + \omega^2 C^2 R^2]$$

$$+ \omega^2 L^2 [(1 - \omega^2 LC)^2 + \omega^2 C^2 R^2]$$

therefore,

$$Z = \left[\frac{R^2 + \omega^2 L^2}{(1-\omega^2 LC)^2 + \omega^2 C^2 R^2}\right]^{1/2} \qquad ...(1.63)$$

Condition for Unity Power Factor

We know that the current in the circuit will be in phase with the applied voltage *if the impedance of the combination is purely resistive* which corresponds when the power *factor is unity.* Then by eq. (1.60), the apparent reactance L_0 must be zero for impedance to be purely resistive, thus from Eq. (1.62),

$$L - C(\omega^2 L^2 + R^2) = 0$$

or $$R^2 + \omega^2 L^2 = L/C \qquad ...(1.64)$$

And the value of angular frequency which makes the apparent reactance zero is denoted by wr given by,

$$\omega_r = 2\pi f_r = \sqrt{\frac{1}{LC} - \frac{R^2}{L^2}} \qquad ...(1.65)$$

or $$f_r = \frac{1}{2\pi}\sqrt{\frac{1}{LC} - \frac{R^2}{L^2}} \qquad ...(1.66)$$

This frequency fr for which power factor is unity is called the resonant frequency for the given parallel circuit corresponding to this frequency the impedance is maximum (but not infinite). The current will be minimum corresponding to this circuit will work as rejector. The variation of current with frequency in parallel resonant is shown in Fig. 1.15 (b). Clearly the current is minimum at resonant frequency. The frequency at parallel resonance given by Eq. (1.66) reduces to the frequency at series resonance given by eq. (1.48) if R is very small so that R^2/L^2 becomes negligible as compared to 1/LC.

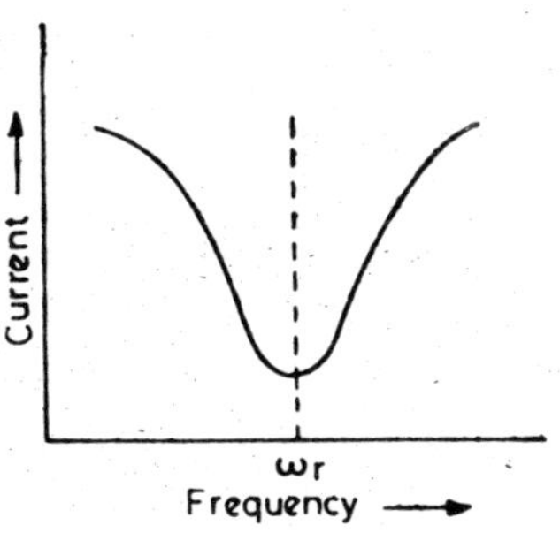

Fig. 1.15(b)

At resonance the impedance of the circuit is obtained by putting the Eqs. (1.64) and (1.65) in (1.63),

$$Z_{Tr} = v\sqrt{\frac{L/C}{\left[1-\left(\frac{1}{LC}-\frac{R^2}{L^2}\right)LC\right]^2+\left(\frac{1}{LC}-\frac{R^2}{L^2}\right)C^2R^2}}$$

$$= \sqrt{\frac{L/C}{\left[1-1+\frac{R^2C}{L}\right]^2+\left(\frac{1}{LC}-\frac{R^2}{L^2}\right)C^2R^2}}$$

$$= \sqrt{\frac{L/C}{\frac{R^4C^2}{L^2}+\frac{C^2R^2}{LC}-\frac{R^4C^2}{L^2}}}$$

$$= \sqrt{\frac{L^2}{C^2R^2}}$$

$$= \frac{L}{CR} \qquad \text{...(1.67)}$$

Dimensionally $\left|\frac{L}{R}\right|$ = [T], time constant

and similarly [CR] = [T]. Therefore $\left[\frac{L}{CR}\right]$ = [R]

is purely resistive and is known as the dynamic resistance of the combination.

Be careful to note that the *resonant frequency in parallel resonance is no longer independent of the resistance in the circuit as in the case with series resonance.*

For parallel resonance to occur,

$$\sqrt{\frac{1}{LC}-\frac{R^2}{L^2}} \text{ should be real}$$

i.e., $$\frac{1}{LC}-\frac{R^2}{L^2} > 0$$

or $$R^2 < L/C$$

or $$R < \sqrt{L/C}$$

Hence the resistance R should be kept as low as possible.

Current Magnification

The A.C. supply current at resonance is obtained by dividing the supply voltage by impedance at resonance *i.e.*,

$$I_r = \frac{\xi_0 e^{j\omega t}}{Z_{TR}} \text{ substituting } Z_{TR} \text{ from eqn. (1.67)}$$

$$= \frac{\xi_0 e^{j\omega t}}{\frac{1}{CR}} = \frac{\xi_0 CR}{L} = I_{ro} \qquad ...(1.68)$$

while the oscillating current J_{Cr} in the capacitor is given by,

$$I_{Cr} = \frac{\xi_0 e^{j\omega t}}{\text{Capacitative reac tan ce}}$$

$$= \frac{\xi_0 e^{j\omega t}}{1/j\omega_r C} = j\xi_0 \xi\omega_r C e^{j\omega t}$$

$$= \xi_0\omega_r C e^{j(\omega t + \pi/2)} \text{ (Since } j = e^{j\pi/2})$$

$$= I_{Cro} e^{j(\omega t + \pi/2)} \qquad ...(1.69)$$

where $I_{Cro} = \xi_0 \omega_r C$

The current magnification is defined as follows:

$$m = \frac{\text{amplitude of current across capaci tan ce}}{\text{amplitude of sup ply current}}$$

$$= \frac{I_{Cro}}{I_{ro}}$$

$$= \frac{\xi_0\omega_r C}{\frac{\xi_0 CR}{L}}$$

$$= \frac{\omega_r L}{R} = Q \text{ (Quality factor by Eq. (1.56a)} \qquad ...(1.70)$$

Similarly the oscillatory current I_{Lr} in the inductor is given by,

$$I_{Lr} = \frac{\xi_0 e^{j\omega t}}{\text{Inductive reac tan ce}}$$

$$= \frac{\xi_0 e^{j\omega t}}{j\omega_r L}$$

$$= -\frac{j\xi_0 e^{j\varepsilon t}}{\omega_r L}$$

$$= \frac{\xi_0}{\omega_r L} e^{j(\omega t - \pi/2)} \quad (\text{as } -j = e^{-j\pi/2})$$

$$= I_{Lro} e^{j(\omega t - \pi/2)}$$

Magnification,

$$m = \frac{I_{Lro}}{I_{ro}}$$

$$= \frac{\xi_0}{\omega_r L} / \frac{\xi_0 CR}{L}$$

$$= \frac{1}{\omega_r CR} = Q \qquad ...(1.71)$$

Thus a parallel circuit (*i.e., rejector circuit), gives a current magnification equal to the quality factor at resonance similar to the voltage magnification by a series (i.e., acceptor circuit)*

Selectivity of a Parallel Resonant Circuit

We have interpreted eq. (1.59) and concluded that impedance variation with frequency in parallel resonance case has exactly the same shape as the current variation with frequency in series resonance case. Fig. 1.15 (c) shows such variation. At resonance the impedance is given by eq. (1.67), *i.e.,*

$$Z_{Tr} = L/CR$$

By *selectivity of the circuit we mean its ability to present a high impedance at resonant frequency and much lower impedance at other frequencies.* Thus, greater the sharpness (or peakness) of the curve (*i.e.,* high Q) the greater is the selectivity. The impedance variation of the parallel circuit is shown in Fig. 1.15 (c),

$$\frac{1}{Z_T} = \frac{1}{Z_L} + \frac{1}{Z_C}$$

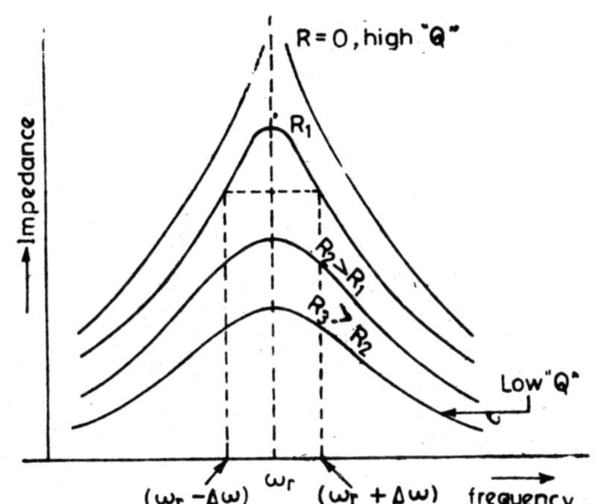

Fig. 1.15(c)

$$= \frac{1}{R + j\omega L} + j\omega C$$

$$= \frac{R + j\omega CR - \omega^2 LC}{R + j\omega L}$$

or $$Z_T = \frac{R + j\omega L}{(1 - \omega^2 LC + j\omega CR}$$

In practice $R \to 0$ and we have,

$$Z_T = \frac{j\omega L}{(1 - \omega^2 LC) + j\omega CR}$$

And its magnitude directly

$$Z_T = \frac{\omega_r L}{\sqrt{(1 - \omega^2 LC)^2 + \omega^2 C^2 R^2}}$$

at $\omega = \omega_r$, $Z_T = Z_{Tr}$

Now if $\omega = \omega_r \pm \nabla\omega$, the terms ωL and $\omega^2 C^2 R^2$ will not differ appreciably from $\omega_r L$ and $\omega_r^2 C^2 R^2$ but the term $(1 - \omega^2 LC)$ will give a considerable change.

Thus when $\omega = \omega_r$ and $R \to 0$, we get,

$$Z_T = \frac{\omega_r L}{\sqrt{(1-\omega^2 LC)^2 + \omega_r^2 C^2 R^2}}$$

It at $\omega = \omega_r + \nabla\omega$

the impedance Z_T is $L/\sqrt{2}$ times its maximum value. Thus,

$$Z_T = \frac{1}{\sqrt{2}} Z_T = \frac{1}{\sqrt{2}} - \frac{L}{CR}$$

At resonance and under condition R → O by Eq. (1.65),

$$\omega_r = 1/\sqrt{LC}$$

$$\frac{1}{\sqrt{2}} - \frac{L}{CR} = \frac{\omega_r L}{\sqrt{[1-(\omega_r + \nabla\omega)^2 LC]^2 + \omega_r^2 C^2 R^2}}$$

On squaring, we get,

$$[1 - (\omega_r + \nabla\omega)^2 LC]^2 + \omega_r^2 C^2 R^2 = 2\omega_r^2 C^2 R^2$$

or $$[1 - (\omega_r + \nabla\omega)^2 LC]^2 = \omega_r^2 C^2 R^2$$

or $$[1 - (\omega_r^2 + \nabla\omega^2 + 2\omega_r \nabla\omega) LC]^2 = \omega_r^2 C^2 R^2$$

The term $\nabla\omega^2$ is very small and neglected,

$$[1 - \omega_r^2 LC - 2\omega_r \nabla\omega LC]^2 = w_r^2 C^2 R^2$$

but $\omega_r^2 = 1/LC \left(\frac{2\nabla\omega}{\omega_r}\right)^2 = \omega_r^2 C^2 R^2$

or $$= \frac{2\nabla\omega}{\omega_r} = -\omega_r CR$$

or $$= \frac{\omega_r}{2\nabla\omega} = \frac{1}{\omega_r CR} = Q$$

$$= \frac{\omega_r L}{R} = \frac{1}{R}\sqrt{L/C} \qquad \text{...(1.72)}$$

$2\nabla\omega$ is the band width of the resonance curve. Thus greater the value of Q corresponds to smaller band width the sharper is resonance and greater selectivity.

COMPARATIVE STUDY OF A SERIES RESONANT AND A PARALLEL RESONANT

It is of interest to compare the two circuit. The Table below gives the result discussed in the last two articles.

Series Resonant Circuit

1. Series resonant frequency is

$$f_r = \frac{1}{2\pi\sqrt{LC}}$$

2. At resonance the power factor is unity and the impedance is purely resistive

$$Z_r = R$$

3. At resonance, the impedance of the circuit is minimum while the admittance is maximum.

4. Series circuit admits *maximum current* at resonant frequency.

5. Series circuit is called *acceptor* circuit because it accepts a particular frequency and rejects others.

6. At resonance the circuit exhibits a *voltage magnification* equal to the quality factor.

Parallel Resonant Circuit

Here, $$f_r = \frac{1}{2\pi}\sqrt{\frac{1}{LC} - \frac{R^2}{L^2}}$$

In this case the power factor is also unity at resonance frequency but the impedance is given by,

$$Z_r = \frac{L}{CR}$$

Here at resonance the impedance of the circuit is *maximum* while the admittance is minimum.

On the other hand this circuit allows a *minimum current* at the resonant frequency.

This circuit is called a *rejector* because it reject only one frequency and accepts others.

This circuit shows a similar *current magnification* equal to the quality factor.

POWER IN A.C. CIRCUIT

power is the rate of doing work. As in an A.C. circuit both the applied emf and the current vary continuously with time, hence the power in an A.C. circuit is equal to the product of the instantaneous emf and instantaneous current averaged over a complete cycle.

The power cannot be calculated by complex number analysis *i.e.*, by directly complex ξ and complex I because the real or imaginary parts of ξI is not equal to the product of the real of imaginary parts of ξ and I respectively. Therefore the power is calculated by actual emf and actual current that is by taking imaginary parts of the complex quantities.

1. Purely Resistance Circuit

In such circuit the current and the applied emf are in the same phase and can be expressed as,

$$\xi = \xi_o \sin \omega t; \quad I = I, \sin \omega t$$

The power $= \xi I = \xi_o I_o \sin^2 \omega t$

Average rate of doing work or power over one cycle,

$$= \frac{\xi_o I_o \int_0^{2\pi} \sin^2 \propto dx}{\int_0^{2\pi} d \propto}. \text{ where } \propto = \omega t$$

$$= \frac{\xi_o I_o}{2\pi} \frac{1}{2} \int_0^{2\pi} (1 - \cos 2 \propto) d \propto$$

$$= \frac{\xi_o I_o}{2} = \frac{\xi_o}{\sqrt{2}} \cdot \frac{I_o}{\sqrt{2}} = \xi_{rms} \times I_{rms} \qquad ...(1.73)$$

Unit of Power

When emf is expressed in volt and current in ampere then unit of power is watt.

2. Power in Inductive Circuit

As we have discussed that the current through a pure inductance lags behind the applied emf in phase by $\pi/2$, therefore,

$$\xi = \xi_o \sin \omega t$$

and $$I = I_o \sin (\omega t - \pi/2)$$

Instantaneous power $= xI$

$$= \xi_o I_o \sin \omega t \sin (\omega t - \pi/2)$$

$$= -\xi_o I_o \sin \omega t \cos \omega t$$

$$= -\frac{\xi_o I_o}{2} \sin 2\omega t$$

Average power over one cycle,

$$= -\frac{\xi_o I_o}{2} \frac{\int_0^{2\pi} \sin 2\propto d\propto}{\int_0^{2\pi} d\propto}$$

$$= \frac{\xi_o I_o}{4\pi} \frac{1}{2} |\cos 2\propto|_0^{2\pi} = 0$$

Thus power dissipation in a pure inductance is zero and the current in such circuit is known as idle or *wattles current.* The choke coil works on this principle.

3. Purely Capacitative Circuit

As we have discussed that the current in a purely capacitative circuit leads the applied emf in phase by $\pi/2$. Thus,

$$\xi = \xi_o \sin \omega t$$

then $\quad I = I_o \sin(\omega t + \pi/2)$

Instantaneous power $= \xi I = \xi_o I_o \sin \omega t \cos \omega t$

$$= \frac{\xi_o I_o}{2} \sin 2\omega t$$

As the average of sin 2 ωt over one cycle is zero. Hence, the *average power is zero.* Thus the average power dissipation in a pure capacitative circuit is zero, hence the current through it is also *wattles or idle.*

4. Circuit Containing L, C and R

As we have discussed that the emf and current are not in phase. These can be written as,

$$\xi\omega = \xi_o \sin \omega t$$

and $\quad I = I_o \sin(\omega t + \theta)$

where phase angle θ is given by,

$$\theta = \tan^{-1} \frac{\omega L - 1/\omega C}{R}$$

Power at any instant

$= \xi I$

$= \xi_o . I_o \sin \omega t \sin (\omega t + \theta)$

$= \xi_o I_o \sin \propto \sin (\propto + \theta), \propto = \omega t$

$= \xi_o I_o [\sin^2 \propto \cos \theta + \sin \propto \cos \propto \sin \theta]$

$$= \xi_o I_o \left[\sin^2 \propto \cos\theta + \frac{1}{2} \sin 2 \propto \sin\theta\right]$$

The average rate of doing work or power over one cycle

$$= \frac{\xi_o I_o \cos\theta \int_0^{2\pi} \sin^2 \propto d \propto}{\int_0^{2\pi} d \propto} + \frac{\frac{1}{2}\xi_o I_o \sin\theta \int_0^{2\pi} \sin 2 \propto d \propto}{\int_0^{2\pi} d \propto}$$

As we have evaluated above that average of $\sin^2\propto = 1/2$ and that of $\sin 2\propto$ is zero. Thus, the second part does not contribute to the power is wattles, hence

Average power over one cycle

$$= \frac{1}{2} \xi_o I_o \cos \theta$$

$$= \frac{\xi_o}{\sqrt{2}} \cdot \frac{I_o}{\sqrt{2}} \cdot \cos\theta$$

$$= \xi_{rms} . I_{rms} \cos \theta \qquad \text{...(1.74)}$$

The term ξ_{rms}. Irms is called apparent power and cos θ is called power factor. Thus,

$$\text{power factor } (\cos \theta) = \frac{\text{True power}}{\text{Apparent power}} \qquad \text{...(1.75)}$$

(i) Power factor in L – C – R circuit:

$$\text{Cos q} = \frac{R}{\sqrt{R^2 + \left(\omega L - \frac{1}{\omega C}\right)^2}} \qquad \text{...(1.76)}$$

(ii) Power factor in inductive circuit:

$$\text{Cos } \theta = \frac{R}{\sqrt{R^2 + \omega^2 L^2}} \qquad \text{...(1.77)}$$

(iii) Power factor in capacitative circuit:

$$\text{Cos } \theta = \frac{R}{\sqrt{R^2 + 1/\omega^2 C^2}} \qquad ...(1.78)$$

CHOKE COIL

As we have discussed the current lags in phase behind $\pi/2$ in a pure inductor leads in Phase by $\pi/2$ in a capacitor when connected in an A.C. circuit. Thus, there is no power loss, as the current through them is *wattless*. Hence, current in an A.C. circuit is reduced by the inclusion of a capacitor or inductor or both with- out much loss of energy. *The inductor which is used for reducing the current in the A.C. circuit is called a choke or choking coil.* A choke coil consists of thick copper wire wound closely in a large number of turns over a soft iron laminated core. The resistance of the choke coil is almost negligible, since loss of power is equal to I^2R. Thus, the loss of power is least, though there is some loss of power due to hysteresis which is much less than P R.

Further, the average power dissipated in the choke coil is given by,

$$P = \frac{1}{2} \xi_o I_o \cos \theta$$

where the power factor,

$$\cos \theta = \frac{R}{\sqrt{R^2 + \omega^2 L^2}}$$

In choke coil, L is large while R is very small. Hence cos θ is nearly zero. Therefore, the power consumed by the coil is extremely small.

Preference of a Choice Coil Over a Resistance

If resistance is used in place of inductance, the power loss is I^2R. Hence, to avoid this loss, we prefer choke coil.

SOLVED EXAMPLES

Example 1:

An alternative valtage of 110 V 50 cycle is applied to a circuit which contains an inductance of 0.02 henry and a resistance of 10 ohms in series. Determine the current and the phase lag.

Solution:

By Eq. (1.39)

$$I_o = \frac{\xi_o}{\sqrt{R^2 + \omega^2 L^2}} = \frac{110}{\sqrt{10^2 + (2 \times 3.1416 \times 50 \times 0.02)^2}}$$

$= 10.5$ amp.

$$\theta = \tan^{-1} \frac{\omega L}{R} = \tan{-1} \frac{2 \times \pi \times 50 \times 0.02}{10}$$

$$= \tan^{-1} \frac{\pi}{5}$$

$= 32°8'$ lagging.

Example 2:

A circuit having a pure inducatance of 0.01 henry in series with a resistance of 5 ohms is fed by 200 V. 50 c/sec. a.c. supply. Calculate the current and its phase, the drop of potential across the inductor and resistor, and the power.

Solution:

$$I_o = \frac{\xi_o}{\sqrt{R^2 + \omega^2 L^2}} = \frac{200}{\sqrt{25 + (2\pi f L)^2}} \text{ on substitution}$$

$$= \frac{200}{\sqrt{25 + (3.14)^2}} = \frac{200}{5.91} = 33.9 \text{ amps.}$$

The angle of lag of current is given by,

$$\theta = \tan^{-1} \frac{\omega L}{R} = \tan{-1} \left(\frac{3.14}{5} \right) = 32°$$

Now the potential difference across the resistance and the inductance are,

$$\xi_R = I_o R = 33.9 \times 5 = 169.5 \text{ volts}$$

and $\xi_L = I_o \omega L = 33.9 \times 3.14 = 106$ volts.

Power dissipation is

$$P = I_o^2 R = (33.9)2 \times 5 = 5746 \text{ volts.}$$

Example 3:

An electric lamp marked 100 volts. d.c. consumes a current of 10 amperes. It is connected to a 200 volts 50 cycles a.c. mains. Calculate the inducatance of the required whole.

Solution:

Given $I_o = 10$ amp. $R = \frac{V}{I} = \frac{100}{10} = 10$ ohms

$\xi_o = 200$ volts, $f = 50$ c/sec.

$\omega = 2\pi f = 2\pi \times 50 = 100\pi$

$$\text{Now } I_o = \frac{\xi_o}{\sqrt{R^2 + \omega^2 L^2}}$$

$$\text{or} \quad 10 = \frac{200}{\sqrt{(10)^2 + (100\pi L)^2}}$$

or $\quad (100\ \pi L)^2 = 300$

therefore,

$$L = \frac{\sqrt{300}}{100\pi} = 0.055 \text{ henry}$$

Example 4:

Find the capacity of a condenser to run a 30 volt, 10 watt lamp when connected series with an alternating e.m.f. of 220 volts and frequency 50 cylces per second.

Solution:

Here, power of the lamp = $V \times I = 10$ watts, $V = 30$ volts.

therefore.

$$I = \frac{W}{V} = \frac{10}{30} = \frac{1}{3} \text{ amp.}$$

$$\text{and} \quad R = \frac{V}{I} = \frac{30}{1/3} = 90 \text{ ohms.}$$

$$\text{Now } I_o = \frac{\xi_o}{\sqrt{R^2 + \left(\frac{1}{\omega C}\right)^2}}$$

$$\text{or} \quad \frac{1}{3} = \frac{220}{\sqrt{(90)^2 + \left(\frac{1}{100\pi C}\right)^2}}$$

On solving, we get, $C = 4.866 \times 10^{-6} F = 4.866 \mu F$

Example 5:

An alternating emf of 200 volts 50 cycles per sec. is applied to a condenser in series with a 100 volts 50 watt lamp. Find the capacity of the condenser.

Solution:

$$\xi I = 50$$

$$I = 50/\xi = 50/100 = \frac{1}{2} \text{ amp.}$$

$$R = \frac{\xi}{I} = \frac{100}{1/2} = 200 \text{ ohms}$$

$$\text{and } R^2 + \frac{1}{\omega^2 C^2} = \left(\frac{\xi}{I}\right)^2 = \left[\frac{200}{1/2}\right]^2 = 400^2$$

$$\frac{1}{\omega^2 C^2} = 400^2 - 200^2 = 12 \times 10^4$$

$$\omega^2 C^2 = \frac{1}{12 \times 10^4}$$

$$\omega = 2\pi f = 2\pi \times 50 = 100\pi$$

$$C = \frac{1}{100\pi \times 100 \times \sqrt{12}}$$

$$= 9.16 \times 10^{-6} \text{ farads}$$

$$= 9.16 \text{ microfarads}$$

Example 6:

A circuit consists of a resistance and capacitance in series. If an alternating emf of 180 volts and 100 c/sec frequency is applied to it, calculate the value of resistance and capacitance when the maximum current is 6 amps. and active power is 360 watts.

Solution:

The circuit resistance is given by power relation

$$R = \frac{P}{I^2} = \frac{360}{6^2} = 10 \text{ ohms.}$$

and the circuit impedance is

$$Z = \frac{V}{I} = \frac{180}{6} = 30 \text{ ohms.}$$

The capacitive reactance is

$$X_c = \sqrt{Z^2 - R^2} = \sqrt{30^2 - 10^2} = 28.3 \text{ ohms.}$$

therefore, the capacitance is

$$C = \frac{1}{\omega X_c} = \frac{1}{2 \times 3.14 \times 100 \times 28.3}$$

= 26.2 micro farad.

Example 7:

A resistance of 10 ohms is joined in series with an inductance of 0.5 henry. What capacitance must be put in series with the combination to obtain maximum current? What will be the potential drop across each element of circuit, if it is connected to 200 V 50 H_x mains?

Solution:

Let the capacity be C, then $\left(\frac{1}{\omega C} = \omega L\right)$ of maximum current which corresponds to resonance

or $$C = \frac{1}{\omega^2 L} = \frac{10^6}{0.5 \times (2 \times 3.14 \times 50)^2} \text{ microfarad}$$

= 20.25 microfarad

Current $I = \frac{V}{R} = \frac{200}{10} = 20$ ampere

potential difference across R = RI = 10 × 20 = 200 V.

Potential difference across L or C = ωLI

= 0.5 × 2 × 3.14 × 50 × 20 = 3142 volts.

Example 8:

A 48 ohms resistor is connected in series with an inductance of 450 mH and a capacitance of 9 microfarad. (i) What is the resonant frequency? (ii) What will be the current in the circuit when it is supplied from a 120 V source operating at resonant frequency? (iii) What is the voltage across the inductance under these conditions?

Solution:

ωL = 1/ωC (condition of resonant)

$$\omega = \frac{1}{\sqrt{LC}} = \frac{1}{\sqrt{450\times10^{-3}\times9\times10^{-6}}} = 2\pi f$$

(i) $f = \dfrac{1}{2\pi\sqrt{450\times9\times10^{-9}}} = 79$ cycles/sec.

(ii) $I_{max} = \dfrac{120}{48} = 2.5$ amp.

voltage across inductance $= \omega L\ I_{max} = 2\pi \times 79 \times 450 \times 10^{-3} \times 2.5$

$= 557.0$ volts.

Example 9:

In a series L.C.R. circuit the values of the various quantities are as follows:

$\xi = 100$ volts, $f_o = 60$ c/sec. $R = 1$ ohm,

$L = 1$ henry and $C = 7.C4$ microfarad.

Prove that the voltage magnification is equal to quality factor Q.

Solution:

Inductive reactance $X_L = \omega_o L = 2\pi r_o L$

$= 2 \times 3.14 \times 60$

$= 377$ ohms (approx.)

Capacitive reactance

$= X_C = \dfrac{1}{\omega_o C} = \dfrac{1}{2\pi f_o C} = 377$ ohms (approx).

At resonance maximum current,

$$I_{max} = \frac{E}{\sqrt{R^2 + (X_L - X_C)^2}} = \frac{E}{R} = 100 \text{ amp.}$$

Therefore potential difference across resonance resistance

$\xi R = I_{max}\ R = 100$ volts.

Potential difference across inductance

= potential difference across capacitance

$= \omega_o L\ I_{max} = \dfrac{I_{max}}{\omega C} = 377 \times 100 = 37700$ volts

$$= Q\xi \left(\text{as} Q = \frac{\omega_0 L}{R} = \frac{1}{\omega_0 CR} = 377 \right)$$

therefore,

Voltage magnification is equal to quality factor Q.

EXERCISES

1. Define complex number. .Explain why complex numbers can be used to express sinusoidal quantities.
2. Define pure inductance. Show that the current in an inductive circuit lags behind the voltage by 90°. Plot the current, self-induced emf and voltage curve and construct the vector diagram for an inductive circuit.
3. Why is a capacitor of practically infinite d.c. resistance? Explain the passage of alternating current through a purely capacitive circuit.

   Deduce equation for current in a capacitive circuit. Plot the current and voltage curves for the capacitive circuit.
4. Explain the impedance and reactance in an a.c. circuit. Distinguish between the two.
5. Write the equations describing the impedance and reactance for a capacitive, inductive and for the following series circuit:
   1. R—L
   2. K—C
   3. R—L—C.
6. Derive expressions for current and impedance when an alternating emf is applied to a circuit having (i) capacity and resistance in series (ii) inductance and resistance in series.
7. Derive an expression for the amplitude and phase of the current in a series circuit consisting of a resistance R, a self-inductance L and a capacitance C, when an alternating emf is applied.
8. Describe the behaviour of a series R—L—C circuit for the case when $X_C > X_L$ and $X_L < X_C$. Discuss the phenomenon of resonance in series R—L—C circuit.
9. What do you understand by the quality factor (Q) of a circuit?

Explain the dependence of the sharpness of the resonance on Q in series L—C—R circuit.

10. Derive an expression for Q in terms of bandwidth of the curve.
11. Prove that Q is the factor by which the applied voltage is magnified across L and C at resonance. Why LCR circuit is often called an acceptor circuit!
12. What is the condition of resonance when an inductor is shunted by a capacitor and explain the paradox of current flowing in the coil and condenser while no current is being supplied by the source at resonance ?
13. What is the difference between parallel resonance and anti-resonance? Derive an expression for impedance when an inductive coil is shunted by a condenser.
14. What do you understand by wattless and power components of an alternating current ?
15. Find the power in an a.c. circuit having: (i) resistance and capacity (ii) resistance and inductance. What is meant by the power factor of a circuit?

16. A non-inductive resistance of 12 ohms is connected in series with a coil of negligible resistance having an inductance of 0.016 henry. An alternating emf of 130 volts 50 cycles is applied to the circuit. Calculate the current and potential difference across the resistance and the inductance.
17. A circuit contains a resistance of 4 ohms and an inductance of 0.68 henry and an alternating effective emf of 500 volts at a frequency of 120 cycles per second is applied to it. Find the value of the effective current in the circuit and power factor.
18. An alternating emf of 200 volts and 50 periods per second is applied to a condenser in series with a 20 volt, 5 watt lamp. Find the capacity of the condenser required to run the lamp.
19. A resistance of 10 ohms is joined in series with an inductance of 0.5 henry. What capacitance should be put in series with the combination to obtain the maximum current ? What will be the potential difference across the resistance, inductance and capacity? The current is being supplied by 200 volts and 50 cycles mains.

20. Alternating emf of 100 virtual volts is applied to a circuit of resistance 0.5 ohm and inductance 0.01 henry, the frequency being 50 cycles/sec. What is the current and lag in the time between emf and current?

21. A RLC circuit has R = 2 ohms, L= 0.005 henry and C = 600 microfarad. If potential difference of 100 volts (rms) at/frequency 50 is applied, calculate (i) current at resonance (ii) power and (iii) drop in volts along each element of the circuit.

22. A coil of resistance 60 ohms and inductance 3 henry is connected in series with a condenser of 4 microfarad and an a.c. supply of 200 volts and 60 c/s. Calculate

    1. The impedance in the circuit
    2. The phase difference between current and voltage.
    3. The potential difference across inductance coil and
    4. The potential difference across the capacitor.

23. A 60 volt 10 watt lamp is to be run on 100 volt 60 cycle mains.

    1. Calculate the inductance of the choke required.
    2. How much pure resistance would be necessary in the circuit to achieve the same result ?

       What is the disadvantage in this case ?

24. An inductance of S milli henry is connected in parallel with the condenser of capacity 500 microfarad. The two are joined in series with a non-inductive resistance of 4 ohms and a source of alternating emf of 100 volts 50 cycles. Determine/the current in the main circuit. At what frequency is the circuit anti-resonant?

Multiple Choice Questions

1. Impedance is composed of

   (a) Resistance and capacitance

   (b) Resistance and inductance

   (c) Resistance and reactance

   (d) Resistance and inductance or capacitance

2. At resonance, circuit impedance is

   (a) Purely capacitive (b) Purely inductive

   (c) Purely resistive (d) None of these

3. Q factor of a coil is measure of its
   (a) mutual inductance (b) self-inductance
   (c) retentivity (d) selectivity
4. In parallel resonant circuit, at resonance the impedance is
   (a) Minimum
   (b) Maximum
   (c) Equal to difference of inductive and capacitive impedances
   (d) None of these
5. A series resonant circuit offers
   (a) zero impedance
   (b) infinite impedance
   (c) very small finite impedance
   (d) none of these
6. An A.C. wave is known by
   (a) undirectional wave
   (b) bi-directional wave
   (c) without directional wave
   (d) none of the above.
7. Vectors are a short hand for the representation of alternating voltages and currents and their use greatly simplifies the problems A.C. work.
   (a) True (b) False
8. A choke has
   (a) A larger number of turns
   (b) Closely spaced turns
   (c) Necessarily an iron core
   (d) None of these
9. At frequencies above resonant frequency, impedance is
   (a) inductive (b) capacitive
   (c) resistive (d) none of these

2

Steady Current

METAL ELECTRODE IN AN ELECTROLYTE

When an electrode of pure zinc is immersed in an electrolyte, as in Fig. 2.1, the chemical forces cause zinc ions to enter the liquid. Since zinc ions are positively charged, the electrode acquires a negative charge. *This process continues until the electric field due to the static charge on the electrode becomes great enough to counteract the chemical forces. Thus, when the* zinc ions are in electrostatics equilibrium, we have,

$$= \vec{E}_c + \vec{E}_{ch} \ 0. \quad ...(2.1)$$

where $\vec{E}_c$ is the electric field intensity due to the static charge on the electrode and $\vec{E}_{ch}$ is the chemical force per unit charge.

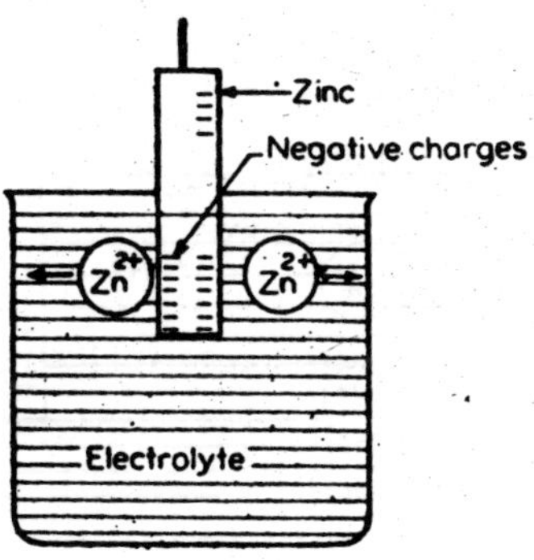

Fig. 2.1

BATTERY ON OPEN CIRCUIT

An electric cell is shown in Fig. 2.2. Positively charged copper ions tend to leave copper electrode just as positively charged zinc ions tent to leave the zinc electrode. The chemical forces may be considered as sum of two electric fields, $\vec{E}_{ch}$ acting on the surface of zinc and $\vec{E}_{ch}$ on the copper surface. $\vec{E}'_{ch}$ is greater in magnitude than $\vec{E}''_{ch}$ and the direction of $\vec{E}'_{ch}$ is opposite to that of $\vec{E}''_{ch}$. Thus, consider $\vec{E}_{ch}$, the chemical force per unit charge to be a vector function of position in the electrolyte, *i.e.,* resultant of these two, we have,

$$\vec{E}_{ch} = \vec{E}'_{ch} + \vec{E}''_{ch} \qquad \text{...(2.2)}$$

The number of zinc ions entering the electrolyte is greater than copper ions. As a consequence the zinc electrode acquires a greater negative free charge than the copper electrode. Hence, *outside of the electrolyte, there is a field $\vec{E}_c$ due to the free charges on zinc and copper electrode $\vec{E}_c$ is directed from copper to zinc electrode.* In the electrolyte, a layer of positive ions surrounds each electrode. The electrode surface and the surrounding ions constitute a double layer of charges, the electrostatic filed intensity $\vec{E}_c$ is directed toward the electrode shown in Fig. 2.2. When equilibrium is established in the double layer, we must have,

$$\vec{E}_c + \vec{E}_{ch} = 0 \qquad \text{...(2.3)}$$

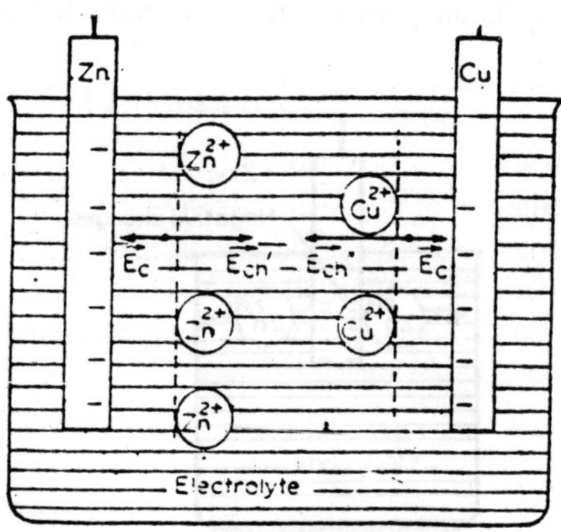

Fig. 2.2

This equation must also be valid for points in the main body of the electrolyte where the ions are in electrostatic equilibrium. Otherwise there would be a continuous migration of ions. Since $\vec{E}_{ch}$ is zero at points relatively far from the electrode, $\vec{E}_c$ must be zero in the main body of electrolyte.

DEFINITION OF EMF

The *electromotive force (EMF) represents the work done by the non-electrostatic (chemical) forces during the transfer of a unit charge from the negative to the positive terminal.* The emf ξ of an electric cell is given by,

$$\xi = \int_m^p \vec{E}_{ch} \cdot d\vec{L} \qquad ...(2.4)$$

Since the electric field, $\vec{E}_{ch}$ exists only in the electrolyte, equation (2.4) may be written, here m denotes minus terminal and p positive,

$$\xi = \oint \vec{E}_{ch} \cdot d\vec{L} \qquad ...(2.5)$$

where the closed path extends from m to p through the electrolyte and then back to p through the space above the electrolyte. *No contribution to the final result is obtained by integrating from p to m because E_{ch} is zero in the space above the electrolyte.* According to equation (2.5), $\vec{E}_{ch}$ is "*non-conservative*", *because the line integral of* $\vec{E}_{ch}$ *around a closed path is not zero.*

DEFINITION OF POTENTIAL DIFFERENCE

Since there are now two electric fields, $\vec{E}_c$ and $\vec{E}_{ch}$, we must re-examine the definition of potential difference. The field $\vec{E}_c$ is due to the time dependent (static) charge distribution in the cell and is, therefore, "*conservative*" *and the potential difference is independent of path between the two points and can be computed only from the conservative electric field.* So the potential difference between the two plates is given by,

$$V_p - V_m = -\int_m^p \vec{E}_c . d\vec{L} \qquad ...(2.6)$$

Since the cell is on open circuit, we have from Eq. (2.3),

$$\vec{E}_c + \vec{E}_{ch} = 0$$

Therefore, substituting for $\vec{E}_c$, we have,

$$V_p - V_m = + \int_m^p \vec{E}_{ch} \cdot d\vec{L} \qquad ...(2.7)$$

Now according to equation (2.4) the right hand side of equation (2.7) is emf. Therefore,

$$V_p - V_m = \int_m^p \vec{E}_{ch} . d\vec{L} = \xi \qquad ...(2.8)$$

Thus, the potential, difference across a cell on open circuit is equal to the emf of the cell.

Since $\oint \vec{E}_c \cdot d\vec{L} = 0$, we can rewrite equation (2.5) as follows,

$$\xi = \oint \left(\vec{E}_c + \vec{E}_{ch}\right) \cdot d\vec{L} \qquad ...(2.9)$$

The emf of an electric cell is equal to the "line integral" of the resultant electric field around a closed path that extends from the minus terminal to the plus terminal through electrolyte and then returns to the minus terminal through the space above the electrolyte.

BATTERY IN CLOSE CIRCUIT

When an electric cell is connected to a resistance as in Fig. 2.3, the negative charge flows through the resistance from the electrode at lower potential (m) to the electrode at higher potential (p). The effect, of course, is the same as that of positive charge flowing from p to m.

Here, for *simplicity and algebraic consistency,* we assume that only positive charge is moving through a wire. When the negative charge on m is reduced. additional positive ions from m enter the solution. This means that in the vicinity of m, $|\vec{E}_{ch}| > |\vec{E}_c|$. When the negative charge on p is increased, practically no positive ions leave p to enter the solution. This means that in the vicinity of p,

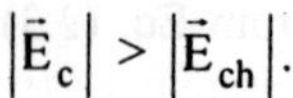

$$|\vec{E}_c| > |\vec{E}_{ch}|.$$

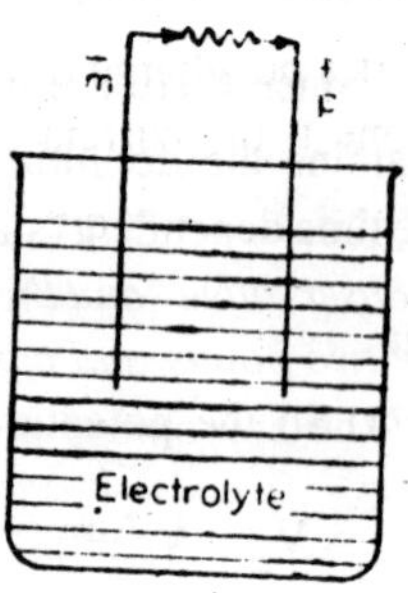

Fig. 2.3

The positive ions diffuse towards p and negative ions towards m. The current in the electrolyte consists of both of these streams of charges. The electric forces are such that positive charges is driven through the electrolyte from m to p. The charge returns to m through the external circuit because of the field $\vec{E}_c$. In a well-constructed cell, this process may continue for a long time.

Now the movement of charge in the electrolyte is related to the electric fields, $\vec{E}_c$ and $\vec{E}_{ch}$. We introduce the symbol $\vec{j}$ called current density which actually related to motion of charge in a conductor is proportional to the field or $\vec{j}$ is function of E,

$$\vec{J} = f(\vec{E}),\ J = 0 = f(0) \qquad ...(2.10)$$

Equation applies to ordinary conductors and excludes "superconductors" for which f(0) is a constant, for many conductors f is linear.

$$\vec{J} = \sigma\vec{E} \qquad ...(2.11)$$

In this case,

$$\vec{J} \propto (\vec{E}_c + \vec{E}_{ch})$$

$$= \sigma(\vec{E}_c + \vec{E}_{ch}) \qquad ...(2.12)$$

where the constant s is called "conductivity" Note that when the cell is disconnected from the resistor, $\vec{J} = 0$, hence, $\vec{E}_c + \vec{E}_{ch} = 0$.

This is same as open circuit case discussed previously.

CURRENT AND CURRENT DENSITY

If a net charge q passes through any cross-section of the conductor in time t, the current i is given by,

$$i = q/t \qquad ...(2.12)$$

Unit

The appropriate RMKS unit is ampere.

$$1 \text{ Ampere} = \frac{1\text{Coul}}{1\text{Sec.}}$$

The current given by equation (2.12) is the constant for the time. If the rate of flow of charge with time is not constant, then i varies with time and is given by,

$$i = \frac{dq}{dt} \qquad ...(2.13)$$

Here we shall consider only constant currents.

Current Density

Current i is a "characteristic" of a particular conductor. Current is a macroscopic quantity, like mass, volume, or length.

A *related* microscopic quantity is the *current density* $\vec{J}$. It is a vector and is *characteristic of a point inside a conductor rather than conductor as a whole.* If the current is distributed uniformly across a conductor or cross-sectional area A, the magnitude of current density for all points on the cross-section is,

$$J = i/A \qquad ...(2.14)$$

If the current is *not uniformly distributed* then equation (2.14) gives the *average density.* However, it is often of interest to consider the *current density at a point. This is defined as the current Δ i through a small area Δ s divided by Δ s, with the limit of this ratio taken as Δ s approaches zero.* Hence, the current density at a point is given by,

$$\vec{J} = \lim_{\Delta s \to 0} \frac{\Delta i}{\Delta s} \qquad ...(2.14a)$$

Unit

Its unit in RMKS is ampere per square metre, as evident from equation (2.14).

The vector $\vec{J}$ at any point is oriented in the direction that a positive charge carrier would move at that point. An electron at that point would move in the direction $-\vec{J}$.

The relationship given in equation (2.14) can be *generalised* for any surface, for a particular surface in a conductor, i is the flux of the vector $\vec{J}$ over that surface or in other words, the *charge flow is represents by a vector $\vec{J}$ which gives the amount of charge passing per unit area per unit time through a surface element at right angles to the flow.*

If we consider the small area Δs shown in Fig. 2.4 at a given place in the material through which the charges are flowing, then amount of charge flowing across that area in a unit time is,

$$\Delta q = \vec{J}.\hat{n}\,\Delta s, \qquad ...(2.15)$$

where $\hat{n}$ is unit vector normal to Ds. Hence, the total charge passing per unit time through any surface,

$$q = \vec{J}.\int_8 \vec{J}.d\vec{S} \qquad ...(2.16)$$

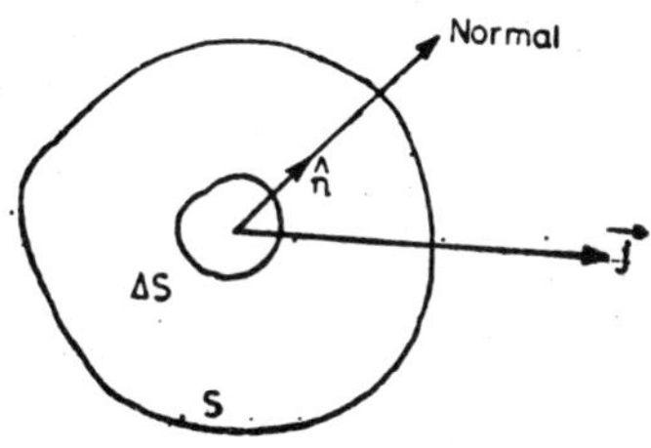

Fig. 2.4

From the definition of the current i = q/t, for unit time *i.e.*, t = 1, therefore,

$$i = q = \vec{J}.\int_8 \vec{J}.d\vec{S} \qquad ...(2.17)$$

DRIFT SPEED (V_D) OF CHARGE CARRIERS IN A CONDUCTOR

The drift speed v_d of charge carriers in a conductor can be computed from the current density $\vec{J}$. Suppose we have a charge distribution shown in Fig. 2.5, the conduction electrons (or charge) are moving to right with the constant drift speed v_d. As this charge distribution passes over the surface elements Δs, the charge Δq flowing across this surface element in a time Δt is equal to the charge contained in a parallelepiped with Δs as its base and height vdΔt. If ρ is the volume density of the charge, then charge contained in the parallelepiped will be given by,

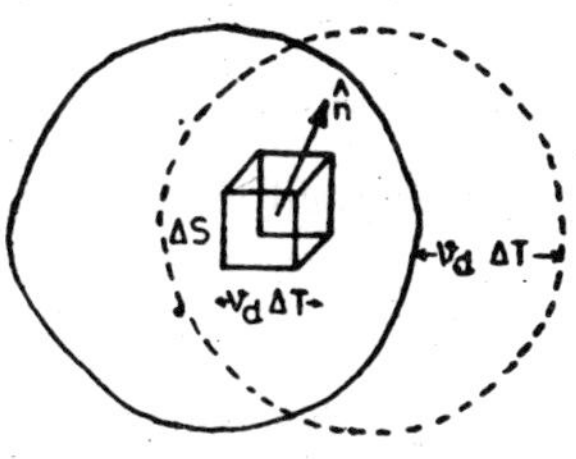

Fig. 2.5

Charge passing through Δs in unit time

= volume of parallelepiped × ρ

= projection of Δs perpendicular to v_d × height of

parallelepiped × ρ

$$\Delta q = \Delta s \cos\theta \, v_d \, \Delta \, t\rho$$

or $$\Delta q = \rho \vec{v}d.\vec{n}dS \, \Delta t$$

Charge per unit time = $\rho \vec{v}d.\vec{n}dS = \rho \vec{v}d.d\vec{S}$ ...(2.18)

Comparing equations (2.18) and (2.16), we get,

$$\vec{J} = \rho \vec{v}d \quad ...(2.19)$$

Let us consider the simple charge distribution of electrons each of charge e and N are the number of electrons per unit volume,

then,

$$r = Ne, \text{ hence, } \vec{J} = Ne\vec{v}d$$

or $$\vec{v}d = \vec{J}/Ne \quad ...(2.20)$$

Example:

Calculate the "drift speed" of electrons shown in Fig. 2.6 and its value when $\vec{J}$ *and* $\hat{n}$ *are parallel.*

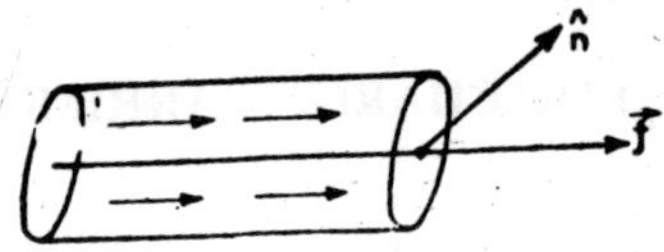

Fig. 2.6

Solution:

We shell use eq. (2.17)

$$i = \int_8 \vec{J}.d\vec{S}$$

Here, the value of integral is independent of the shape of the surface.

and, $i = \vec{J}.\vec{A}$

where, A is the area of cross-section of cylinder.

If area A is normal to the surface,

$$i = J.A, \text{ But } J = Nev_d$$

Hence,

$$i = Nev_dA \text{ or } \quad v_d = \frac{i}{NA_e}$$

CONSERVATION OF CHARGE *i.e.,* CONTINUITY EQUATION

As we know that the electric charge is "*indestructible*", it is never *lost* or *created* that is the charge is conserved. The current i out of a closed surface represents the rate at which the charge leaves the volume v enclosed by 8. If there is a "*net current*" out of a closed surface *i.e.,* the net charge has crossed the surface. *The amount of charge inside must decrease correspondingly according to conservation rule,*

$$i = \int_8 \vec{J}.\hat{n}dS = -\frac{d}{dt}(Q_{in}) \qquad ...(2.21)$$

where Q_{in} is the charge inside the surface. But

$$Q_{in} = \int_v \rho dv$$

therefore,

$$i = \int_S \vec{J}.\hat{n}d\vec{S} = -\frac{d}{dt}\int_v \rho dv$$

using divergence theorem, we get,

$$\int_8 \vec{J}.\hat{n}dS = \int_8 \vec{\nabla}.\hat{n}dS\ dv$$

hence $$\int_v \vec{\nabla}.\vec{J}.dv = -\frac{\delta}{\delta t}\int_v \rho dv$$

i.e., $$\nabla.\vec{J} = -\frac{\partial\rho}{\delta t} \qquad ...(2.22)$$

This is an important equation regarding "*charge flow*", *called* **continuity equation** *and describes the fact that the charge is conserved.* The decrease in value of charge inside a small volume must correspond to a flow of charge.

A steady state is one in which all time derivatives are zero so we must have $\frac{\delta\rho}{\delta t} = 0$ and

$$\vec{\nabla}.\vec{J} = 0 \text{ (Steady state)}$$

or $$\int_8 \vec{J}.d\vec{S} = 0 \qquad ...(2.24)$$

Now, if the medium obeys Ohm's law, we have,

$$\vec{J} = \sigma\vec{E}$$

then we get,

$$\vec{\nabla}.\vec{J} = \sigma\,\vec{\nabla}.\vec{E} = 0$$

So for a steady state we *conclude that E has zero divergence, hence according to Gauss' law, ρ vanishes and Laplace equation is valid.*

Since,

$$\vec{E} = -\vec{\nabla}.V,$$

we have Laplace equation,

$$-\vec{\nabla}^2 V = \vec{\nabla}.\vec{E} = \frac{\rho}{\varepsilon_0} = 0$$

OHM'S LAW

Ohm's law states that the potential difference between the ends of a conductor varies directly as the current flowing in it, provided the temperature does not change and the physical state of the conductor remains the same.

Potential difference ∝ Current

or $$\frac{\text{Potential difference}}{\text{Current}} = \text{Constant}$$

= Resistance of the conductor.

Resistance

By definition, the resistance R_{ab} between the terminals a and b on a conductor is given by,

$$R_{ab} = \frac{V_a - V_b}{i} \qquad ...(2.25)$$

where the current i is directed from a to b. This relation is called Ohm's law.

Unit

Resistance Rab is expressed in ohms

1 ohm = 1 Volt/1 ampere.

The ohm *is defined as the resistance of a conductor in which a potential difference of* 1 *volt is developed when a current of 1 ampere flows through it.*

Actually, *resistance has value beyond that of a mere definition.* For any conductors, the resistance depends only on the temperature and does not depend on the potential difference or the current. In *other words, the resistance is a property of the conductor and is not related to the circuit in which the conductor is connected.*

The equation R = V/i can be used to calculate the current in a conductor if we know the resistance and the potential difference across the conductor.

RESISTANCE OF CYLINDRICAL CONDUCTOR AND SPECIFIC RESISTANCE

An example let us calculate the resistance of the cylindrical conductor of radius r and length l as shown in Fig. 2.7.

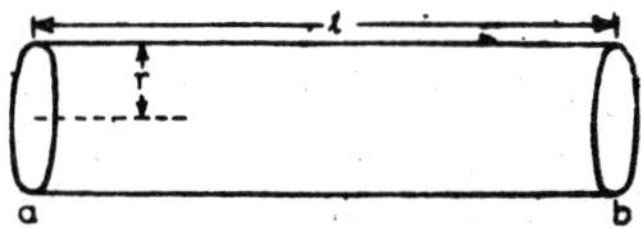

Fig. 2.7

The potential difference,

$$V_a - V_b = -\int_b^a \vec{E}.d\vec{l}$$

Now using $\vec{J} = \sigma\vec{E}$ and reversing the limit,

$$V_a - V_b = \int_a^b \frac{\vec{J}.d\vec{l}}{\sigma}$$

From definition, the magnitude of $J = \frac{i}{A} = \frac{i}{\pi r^2}$.

Hence,

$$V_a - V_b = \int_a^b \frac{idl}{\sigma\pi r^2}, \text{ (Since i and dl are parallel)}$$

$$= \frac{il}{\sigma\pi r^2}$$

Therefore,

$$R = \frac{V_a - V_b}{i} = \frac{l}{\sigma\pi r^2} \qquad ...(2.26)$$

The resistance of a conductor of uniform cross-section is directly proportional to the length and inversely proportional to the conductivity and the cross-sectional area.

Specific Resistance

In relation (2.26), let us take a wire of unit length and unit cross-section, then,

$$R = \frac{1}{\sigma} = \rho_8 \qquad ...(2.27)$$

The quantity ρ_s is called "*specific resistance*" or "*resistivity*". Thus, the *specific resistance is the resistance offered by a conductor of unit length and unit cross-section.* Its unit is ohm-metre. "*Conductance*," the "*reciprocal of resistivity*" is electrical conductivity or simply conductance. Its unit is mho.

OHM'S LAW AT A POINT

Consider a block of conducting material as shown in Fig. 2.8.

Let a small imaginary rectangular shell of length l and cross-section a be constructed around a point P in the interior of the block with a normal to $\vec{J}$ as indicated in Fig. 2.8.

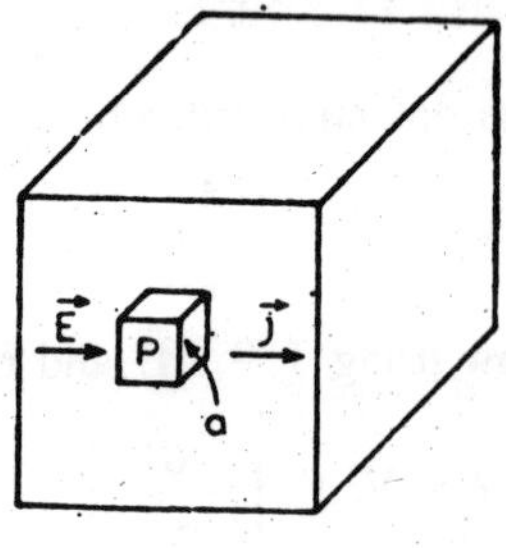

Fig. 2.8

Then, on applying Ohm's law to this shell, we have,

$$V = iR$$

where V = potential difference between ends of the shell. But $V = E_l$, where E is the electric field and $J = i/a$ or $i = Ja$. Therefore,

$$E_l = V = iR = JaR$$

or $$\vec{J} = \frac{1}{aR} E$$

But according to equation (2.26), *i.e.*,

$$R = \frac{l}{\sigma a}$$

since here $\pi r^2 = a$,

or $$\sigma = \frac{l}{aR}$$

Therefore,

$$\vec{J} = \sigma\vec{E} \qquad ...(2.28)$$

By making the shell enclosing P as small as we wish, this relation may be made to apply at the point P and we may write as Eqn. (2.28).

RESISTANCE ANALOGOUS TO RECIPROCAL OF CAPACITANCE

In fact, this is the relationship between "conductivity" and "permittivity". In many respects, the calculation of the resistance between two electrodes is the same as that of capacitance. The similarity may be seen by comparing the definition of R and C, namely,

$$R = \frac{V_a - V_b}{i}, \; C = \frac{V_a - V_b}{i}.$$

The current i is given by,

$$i = \oint_s \vec{J}.d\vec{S}$$

Similarly, the charge q may be written by Gauss's law

$$q = \oint_s \vec{D}.d\vec{S}$$

If the medium has a conductivity σ and permittivity ε, then

$$\vec{J} = \sigma\vec{E} \text{ and } \vec{D} = \varepsilon\,\vec{E}$$

Therefore,

$$i = \oint_s \sigma\vec{E}.d\vec{S} \text{ and } q = \oint_s \varepsilon\vec{E}.d\vec{S}$$

Thus,

$$R = \frac{V_a - V_b}{\oint_s \sigma\vec{E}.d\vec{S}} \text{ and } C = \frac{\varepsilon\oint_s \vec{E}.d\vec{S}}{V_a - V_b}$$

Consequently,

$$RC = \varepsilon/\sigma$$

$$R = \epsilon/\sigma C \qquad ...(2.29)$$

If the capacitance between two conductors separated by a medium of permittivity ε is known, then the resistance between the same two conductors separated by a medium of conductivity σ may be calculated by means of Eq. (2.29).

Example:

As an example, let us calculate the resistance between two parallel plates. The capacitance C between two parallel plates of area A separated by distance l is,

$$C = \varepsilon A/l$$

Solution:

By Eq. (2.29), the resistance,

$$R = \frac{\varepsilon}{\sigma(\varepsilon A / l)} = \frac{l}{\sigma A} = \frac{\rho_8 l}{A} \qquad ...(2.30)$$

Same result as expressed by (2.26), and we can again define a specific resistance.

WIEDMANN AND FRANZ LAW

As we know that good conductors of heat are also the good conductors for electric current. The Wiedmann and Franz *established "empirically" the relation between the thermal and electrical conductivities.* If σ_h is the thermal conductivity then the ratio σh/σ at higher temperature is proportional to the absolute temperature,

$$\frac{\sigma h}{\sigma} = LT \qquad ...(2.31)$$

where L is the constant of proportionality, is constant for all metals and is known as "Lorentz number".

THE RELAXATION TIME

The approach of current in an extended medium to a steady state can be studied by using,

$$\vec{\nabla}.\vec{J} + \frac{\delta\rho}{\delta t} = 0 \text{ and } \vec{J} = \sigma\vec{E}$$

together with $\vec{\nabla}.\vec{E} = \rho/\varepsilon o$, where ρ is charge density and we assume that the medium is conducting but not also polarizable Let us substitute $\vec{J} = \sigma\vec{E}$ in continuity equation, we get,

$$\vec{\nabla} \cdot \sigma \vec{E} + \frac{\delta \rho}{\delta t} = 0$$

or $$\sigma \vec{\nabla} \cdot \vec{E} + \frac{\delta \rho}{\delta t} = 0$$

or $$\frac{\sigma \rho}{\varepsilon_0} + \frac{\delta \rho}{\delta t} = 0, \text{ since } \nabla \cdot \vec{E} = \frac{\rho}{\varepsilon_0}$$

or $$\frac{\delta \rho}{\rho} = -\frac{\sigma dt}{\varepsilon_0}$$

Integrating the above equation,

$$\int \frac{d\rho}{\rho} = \log_e \rho = -\int \frac{\sigma}{\varepsilon_0} dt$$

$$= -\frac{\sigma t}{\varepsilon_0} + \text{Const (a)}$$

or $$\log_e \pi = -\frac{\sigma t}{\varepsilon_0} + a$$

Taking antilog, we get,

$$\rho = \exp\left[-\frac{\sigma t}{\varepsilon_0} + a\right]$$

or $$= e^a \, e^{-\sigma t}/\varepsilon_0$$

e^a can be evaluated by setting $t = 0$, then, $\rho = \rho_0$, the fixed charge, *i.e.,*

$$e^a = \rho_0$$

hence, $\rho = \rho_0 e^{-\sigma t}/\varepsilon_0$ ...(2.32)

We see that the time required for an initial charge density to fall to 1/e of its value is ε_0/σ. It is called "relaxation time".

ELECTRIC POWER AND WORK

An electric generator is shown in Fig. 2.9 connected to an electric heater. Let us calculate the amount of power supplied to the resistance R and the amount of heat developed in the resistor in a time t. The potential difference,

$$V_a - V_b = -\int_b^a \vec{E} . d\vec{l} = \int_a^b \vec{E} . d\vec{l}$$

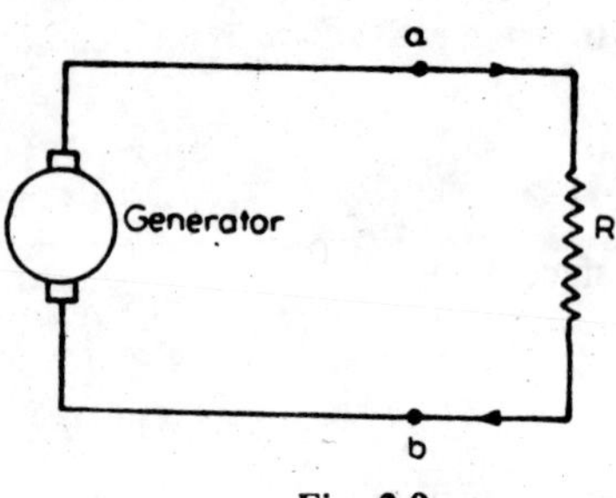

Fig. 2.9

The work done in moving a charge dq from a to b is,

$$dw = (V_a - V_b)\, dq \qquad ...(2.33)$$

(Joules) = (Volts) (Coulombs)

By definition, power P is the rate of doing work. Thus,

$$P = \frac{dw}{dt} = (V_a - V_b) \frac{dq}{dt}$$

or $$P = (V_a - V_b)\, i$$

(Watts) = (Volts) (Amperes)

Since, by Ohm's law,

$$V_a - V_b = iR$$

hence,

P = i2R joules/see or watt

or, $$P = \frac{(V_a Vb)^2}{R} \text{ jouls/see. or watt} \qquad ...(2.34)$$

This is the amount of energy absorbed per second in the resistance R. Total heat developed in R in time T,

$$w = \int_O^T Pdt = \int_O^T e^2 Rdt \qquad ...(2.35)$$

where w is expressed in joules, equation (2.33) indicates that joule = watt. see.

EMF and Internal resistance

If a source of emf is connected in series with resistance R as in Fig. 2.10. If r is the internal resistance of B, the work done by the generator is the work done by the field E causing the charge to move from a to b. Consequently the rate at which the generator does work on the battery is given by,

$$V_b - V_a = -\xi - ir$$

where, – x is the drop in voltage (due to emf) in passing from the plus terminal to the minus terminal and – ir is the drop in voltage in passing through the internal resistance r. The generator maintains the potential difference $V_a - V_b$. The work done by the generator is the work done by the field E in moving charge from a to b. Consequently, the rate at which the generator does work on the battery is,

$$(V_a - V_b)\ i = \xi i + i^2 r$$

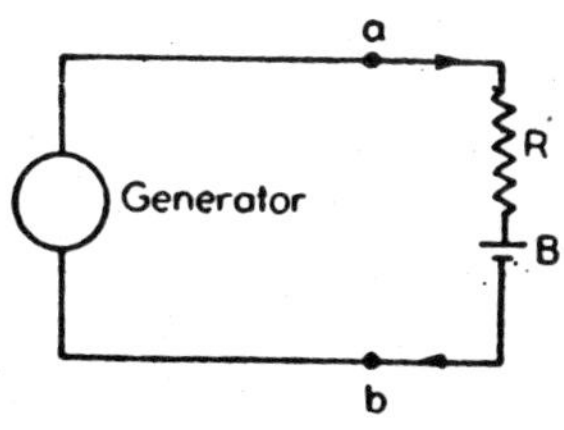

Fig. 2.10

That the power absorbed in the battery due to emf is ξi and the power absorbed in the battery due to the evolution of heat is $i^2 r$.

Finally, let us take the discharge of battery through the resistance as shown in Fig. 2.11.

$$\xi = iR + ir$$

multiply both sides of this equation by i, we get,

$$\xi i = i^2 R + i^2 r$$

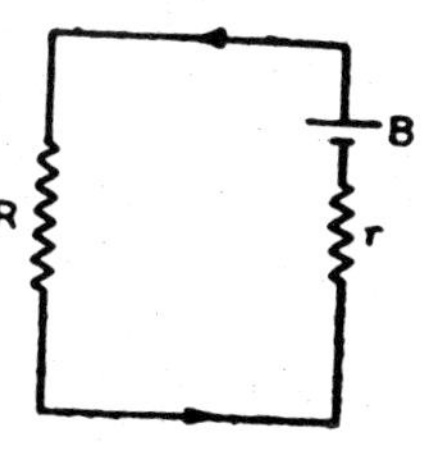

Fig. 2.11

Here ξi is the rate at which work is being done by chemical forces within the battery. This power is equal to the rate at which the heat is developed in the external resistance ($i^2 R$) plus the rate at which heat is developed in the internal resistance ($i^2 r$).

MAXIMUM POWER THEOREM

Let the cell of emf ξ and internal resistance r be connected in series with an external resistance R. Then the current i in the circuit,

$$i = \frac{\xi}{R + r}$$

Power consumed $P = i^2R$

$$= \frac{\xi^2 R}{(R + r)^2}$$

$$= x^2R\ (R + r)^{-2}$$

Differentiating w.r.t. R,

$$\frac{dP}{dR} = x^2\ [(R + r)^{-2} - 2R\ (R + r)^{-3}]$$

$$= x^2\ (R + r)^{-3}\ [R + r - 2R]$$

$$= x^2\ (R + r)^{-3}\ (r - R)$$

$$\frac{dP}{dR} = \frac{\xi^2 (r - R)}{(R + r)^3}$$

Now, $\frac{dP}{dR} = 0$

when $R = r$ or when $R = \infty$. If $R = r$, *the power consumed is maximum and if* $R = \infty$ *power consumed is zero.* This is known as "maximum power theorem" and *sates that the power consumed is maximum in the internal circuit if the external resistance is equal to the resistance of the source of emf.* Theorem is applicable to cells, thermocouples, dynamos, thermionic valves etc.

KIRCHHOFF'S LAWS

Ohm's law is unable to give values of currents in various parts of a complicated circuit. Kirchhoff in 1842 gave two general laws for calculating the currents and resistances in networks at junctions. *These laws are obtained from the law of "conservation of energy".*

1. Kirchhoff's First Law *i.e.,* Current Law

It applies to circuit nodes (or junction) and states *that in any network, the algebraic sum of the currents at any junction in a circuit is zero, i.e.,*

$$\Sigma\ i = 0 \qquad ...(2.37)$$

the currents included in the above sum differ in sign according to whether they flow to or from the junction.

Proof:

We can prove this law as follows. For steady current the equation of continuity is given in equation (2.24), *i.e.*,

$$\int_{s} \vec{J}.d\vec{S} = 0$$

and for a closed surface,

$$\oint \vec{J}.d\vec{S} = 0.$$

This relation can be applied to any volume. For example, the volume may be entirely inside of a conducting medium, or it may be only partially filled with conductors. The conductors may form a network inside the volume, or they may meet at a point.

As an illustration of this latter case, the surface S in Fig. 2.12 encloses a volume that contains five conductors meeting at a junction point P. *Taking the current flowing always from the junction as negative and the current flowing towards the junction as positive,*

according to relation $\oint \vec{J}d\vec{S} = 0.$

we get, $i_1 - i_2 - i_3 + i_5 = 0.$

It follows,

$$i_1 + i_4 = i_2 + i_3 + i_5.$$

In accordance with the above equation, Kirchhoff's current law can be stated in different way, *that the total current entering any junction (or node) of a circuit equals the total current leaving that junction.*

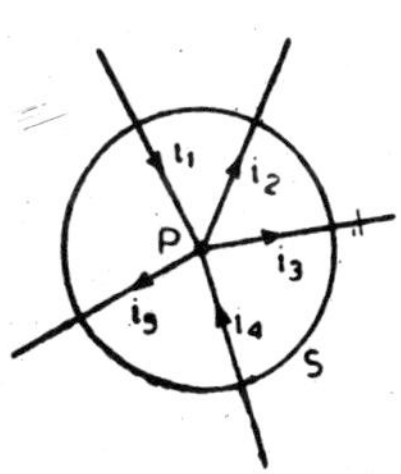

Fig. 2.12

"Physically," Kirchhoff's current law *implies that there can be no "accumulation" of electric charge at any junction of a circuit and is nothing more than a restatement of the principle of conservation of charge.*

2. Kirchhoff's Voltage (Second) Law or Loop Rule

It applies to close circuits (meshes) and states that in any closed circuit, *algebraic sum of the products of the current and resistance of the each part of the circuit is equal to the total emfs acting in the circuit, i.e.,*

$$\Sigma ir = \Sigma \xi \qquad ...(2.38)$$

and if ξ is not acting *i.e.*, if there is no source of emf, then,

$$\Sigma ir = 0$$

Proof:

To prove Kirchhoff's voltage law, let us consider the simple electric circuit shown in Fig. 2.13. The current is i at all points in the circuit, we have from Ohm's law at a point,

$$\vec{J} = \sigma\vec{E}$$

where $\vec{E}$ is total field at the point.

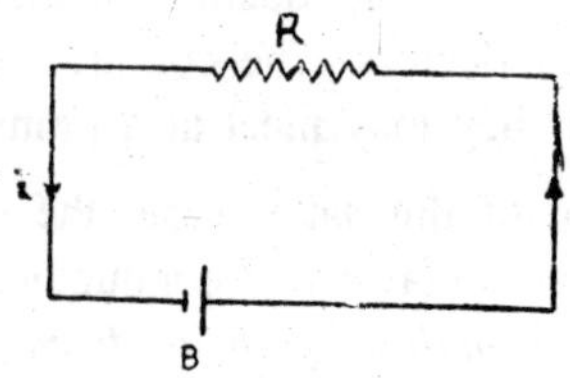

Fig. 2.13

In general the total $\vec{E}$ may not only be due to static charges but also to other causes such as the chemical action in a battery.

We may write,

$$\vec{E} = \vec{E}_c + \vec{E}_{ch}$$

where E_c is field due to static charge on the electrodes and E_{ch} is the field due to chemical forces.

If a is the cross-sectional area of the conductor, then,

$$J = i/a \text{ and } \sigma = l/aR$$

(See equation (2.26), where l is the length of the conductor).

Hence,

$$E = \frac{\vec{J}}{\sigma} = \frac{i}{a} = \frac{aR}{l} = \frac{iR}{l} = \vec{E} = \vec{E}_c + \vec{E}_{ch}$$

Integrating around the complete circuit, we get,

$$\oint \vec{E}_c . d\vec{l} + \oint \vec{E}_{ch} d\vec{l} = i\oint \frac{R}{l} dl$$

$\oint \vec{E}_c . d\vec{l} = 0$, because $\vec{E}_c$ is a conservative field and $\oint \vec{E}_{ch} . d\vec{l}$ is emf ξ according to equation (2.8), and $\oint R / l\, dl = R_T$. Hence,

where R_T is the total resistance and is equal to R if internal resistance is zero. We can generalize for number of closed circuits *i.e.*,

$$\Sigma\xi = \Sigma ir.$$

If ξ were absent, the,

$$\Sigma ir = 0.$$

While applying the second law, the *positive value of the current is taken when we are traversing in the direction of the current and the emf is taken positive when we traverse from the negative to the positive electrode through the electrolyte.* In circuit ABDA in Fig. 2.14,

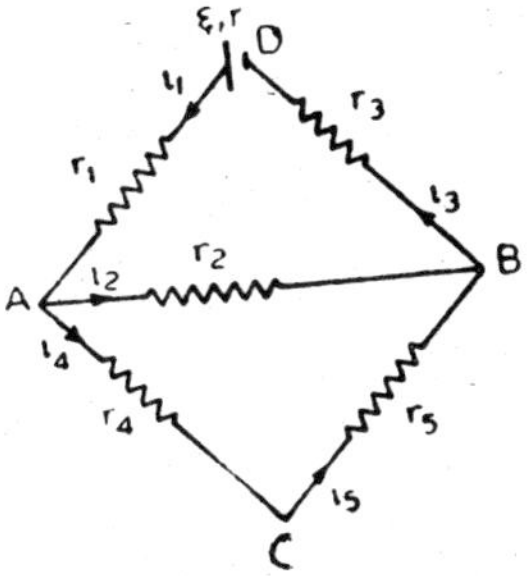

Fig. 2.14

$$i_1 r_1 + i_2 r_2 + i_3 r_3 + ir = x$$

and in circuit ABCA

$$i_2 r_2 - i_5 r_5 - i_4 r_4 = 0.$$

APPLICATION OF KIRCHHOFF'S LAW TO WHEATSTONE'S BRIDGE

Fig. 2.15 represents the Wheatstone's bridge circuit where P, Q, R and S are connected to form a closed circuit. A cell is connected between the points A and C and a galvanometer is connected between the points B and D. The currents through the various branches are indicated in the figure. i_g is the current through galvanometer and G is its resistance. Applying the Kirchhoff's first law, we get, at the junction B,

$$i_1 - i_2 - i_g = 0 \qquad ...(1)$$

at the junction D,

$$i_3 + i_g - i_4 = 0 \qquad ...(2)$$

Now, applying Kirchhoff's second law to the meshes ABDA and ABCDA, we get,

$$i_1 P + i_g G - i_3 R = 0 \qquad ...(3)$$

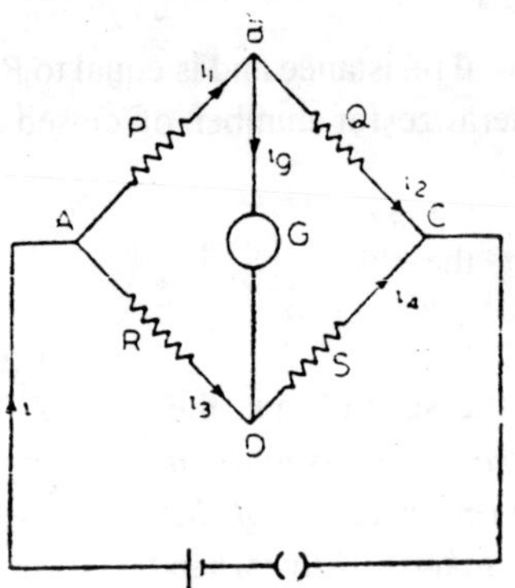

Fig. 2.15

$$i_1P + i_2Q - i_4S - i_3R = 0 \qquad ...(4)$$

When the galvanometer shows, "zero deflection", the points B and D are at the same potential and $i_g = 0$. Hence, equation (1), (2) and (3) reduce to

$$i_1 = i_2 \qquad ...(5)$$

$$i_3 = i_4 \qquad ...(6)$$

$$i_1P = i_3R \qquad ...(7)$$

Putting the values of (5) and (6) in equation (4), we get,

$$i_1P + i_1Q - i_3S - i_3R = 0$$

or $$i_1 (P + Q) = i_3 (R + S) \qquad ...(8)$$

Now divide equation (8) by (7), we get

$$\frac{i_1(P+Q)}{i_1P} = \frac{i_3(R+S)}{i_3R}$$

or $$\frac{P+Q}{P} = \frac{R+S}{R}$$

or $$\frac{Q}{P} = \frac{S}{R}$$

or $$\frac{P}{Q} = \frac{R}{S}. \qquad ...(2.39)$$

Thus, at balance point, *i.e., where the galvanometer shows zero deflection, we have, P/Q = R/S. We can find the value of an unknown resistance S in terms of known resistance P, Q and R.*

Galvanometer current i_g when the Wheatstone bridge is unbalanced. When the bridge is unbalanced, the galvanometer current, i_g can be determined by Kirchhoff's laws.

At the point A Kirchhoff's first law gives,

$$i - i_1 - i_3 = 0$$

or $$i_3 = i - i_1 \quad ...(1)$$

At the point B, we get,

$$i_1 - i_2 - i_g = 0$$

or $$i_2 = i_1 - i_g \quad ...(2)$$

and at the point C, we have,

$$i_2 + i_4 - i = 0$$

or $$i = i_2 + i_4 = i_1 - i_g + i_4$$

or $$i_4 = i - i_1 + i_g \quad ...(3)$$

Applying Kirchhoff's second law to the mesh ABDA, we get,

$$i_1 P + i_g G - i_3 R = 0.$$

Substituting i3 from (1), we get,

$$i_1 P + igG - (i - i1)\ R = 0$$

or $$i_1\ (P + R) + i_g G - iR = 0 \quad ...(4)$$

From the mesh BCDB, we get,

$$i_2 Q - i_4 S - i_g G = 0.$$

Substituting i_2 and i_4 from (2) and (3) in the above equation, we get,

$$(i_1 - i_g)\ Q - (i - i_1 + i_g)\ S - i_g G = 0$$

or $$i_1\ (Q + S) - i_g\ (Q + S + G) - i_S = 0 \quad ...(5)$$

Multiplying equation (4) by (Q + S) and (5) by (P + R), then subtracting resulting (5) from resulting (4), we get,

$$i_g\ [G(Q + S) + (P + R)\ (Q + S + G)] - i[R\ (Q+S) - S(P+R)] = 0$$

or $$i_g\ [G(Q + S) + (P + R)\ (Q + S + G)] - i(QR - SP) = 0$$

or $$i_g = i \frac{QR - SP}{G(P+Q+R+S)+(Q+S)(P+R)} \quad ...(2.4)$$

From the above equation, we can also find the Wheatstone's bridge condition of balance by putting, ig = 0. Then,

$$QR - SP = 0$$

or $$P/Q = R/S$$

which is same as equation (2.39).

SENSITIVITY OF A WHEATSTONE'S BRIDGE

The condition for the balance of Wheatstone's bridge is P/Q = R/S. This corresponds to zero current in the galvanometer. Now, if the bridge is slightly out of balance, there will be a current in the galvanometer, but the current is too feeble, hence cannot be detected easily so that the bridge may not be very sensitive. Thus, the sensitivities of the bridge depends on the sensitivity of the galvanometer.

In Wheatstone bridge, if we interchange the galvanometer and battery, the condition for the null point will not change, however, the two arrangements will not have the same sensitivity. The following discussion about the sensitiveness is due to H.A. Calendar.

The unknown resistance r is connected in the arm CD and i be the current in the arm CD and i be the current in it. Let the resistance in the arm AB, BC, DA and the galvanometer be mmr_1, mr_1, nr_1 and G respectively. Let the current in the galvanometer be i_g and in the arm AB i_1, then the current in the arm AB i_1, then the current in the arm AD will be $i_1 + i_g$ as shown in Fig. 2.16. The bridge balance condition is according to equation (2.39),

$$\frac{mnr_1}{mr_1} = \frac{nr_1}{r}$$

or $$r_1 = r$$

Thus, the bridge condition is satisfied if r1 = r or in other words, if $r = r_1$, the bridge is balanced. Hence, *the difference $r - r_1$ is the measure of want of balance i.e., the greater the difference, the bridge id far from the balance.* However, a bridge will be most sensitive, if the deflection in the galvanometer is large, even when $r - r_1$ is extremely small. Applying Kirchhoff's second law to the circuits ABDA and DCBD,

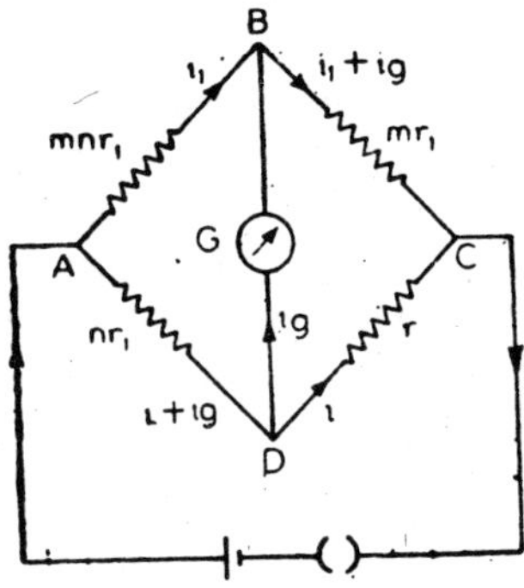

Fig. 2.16

$$i_1 mnr_1 - i_g G - (i + i_g)\, nr_1 = 0$$

and $$ir - (i1 + i_g)\, mr_1 - i_g G = 0.$$

On rewriting the above equations,

$$i_1 mnr_1 - i_g (G + nr_1) - inr_1 = 0 \qquad ...(1)$$

and $$-i_1 mr_1 - i_g (G + mr_1) + ir = 0 \qquad ...(2)$$

Multiplying equation (2) by n and adding the resulting equation to (1), we get,

$$in (r - r_1) - i_g [n(G + mr_1) + (G + nr_1)] = 0$$

or $$\frac{i_g}{i} = \frac{n(r - r_1)}{(1 + n)G + n(1 + m)r_1}$$

or $$\frac{i_g}{i} = \frac{r - r_1}{\left(1 + \frac{1}{n}\right)G + (1 + m)r_1} \qquad ...(2.41)$$

The ratio ig/i depends on the want of balance $r - r_1$, the galvanometer resistance G and the ratio m and n. This ratio is independent of the resistance of the battery.

For greater sensitiveness for a given want of balance, $r - r_1$, the value of i_g should be large, therefore,

(i) i should be large,

(ii) G should be small

(iii) n should be large

(iv) m should be small.

For increasing the sensitivity the current i should be large. But the current i cannot be increased indefinitely as this will heat up the resistance and consequently alter their resistance. Further, the sensitivity increases with decrease in the resistance of the galvanometer, but this decrease in the galvanometer resistance is achieved by decreasing the number of turns in the coil of the galvanometer and which *decreases the sensitivity of the galvanometer itself.*

Therefore, we have to increases sensitivity by increasing n and decreasing m. In the limiting case, if $n = \infty$ and $m = 0$,

$$\frac{i_g}{i} = \frac{r - r_1}{G + r_1} \qquad ...(3)$$

This condition gives the "*ideal sensitivity*" which is *impracticable* because (1) if $n = \infty$, the resistance of the arms AB and AD becomes infinite and (2) if $m = 0$; the points A and C are short circuited. However, making n very large and m very small has no great advantage, because even if $n = m = 1$, the sensitivity from equation (2.41) is

$$\frac{i_g}{i} = \frac{r - r_1}{2(G + r_1)} \qquad ...(4)$$

This value is still half of the value in the ideal limiting case given in equation (3).

Hence, the necessary condition is that $n > 1$ and $m < 1$ *i.e.*, n should not be very small or m too large because in either case the sensitivity will be very much reduced. Thus, *the Calender rule is that the sensitivity of the bridge will be higher if the resistance in series with the unknown resistance is greater than the resistance connected in parallel to it.*

It can be easily shown that for greater sensitivity, the galvanometer should have the same order of the resistance as the unknown resistance.

Another rule for greater sensitivensess is the "greater of the two resistances, the galvanometer or the battery should be connected between the junction of the two highest and two lowest resistances".

MAXWELL'S CYCLIC CURRENTS

Equations obtained by the application of Kirchhoff's law are sometimes very tedious to solve in a complicated network. Maxwell suggested a simple device, called cyclic currents method, which simplifies the solution of problems. According to this method, a certain magnitude of current is assumed to flow in each mesh, all the cyclic currents being in the same direction *i.e.,* either clockwise or anti-clockwise. The actual current flowing in any branch will then be the difference between the cyclic currents of the meshes, if separate.

APPLICATIONS OF MAXWELL'S CYCLIC CURRENT METHOD

1. Wheatstone's Bridge

Let x, y and z be the currents flowing in the clockwise direction through the meshes ABDA, BCDB and ADCA as shown in Fig. 2.17. Applying Kirchhoff's second law, respectively we get,

$$xP + (x - y)\,R_g + (x - z)\,R = 0 \qquad ...(1)$$

$$yQ + (y - z)\,S + (y - x)\,R_g = 0 \qquad ...(2)$$

$$(z - x)\,R + (z - y)\,S + zr = 0 \qquad ...(3)$$

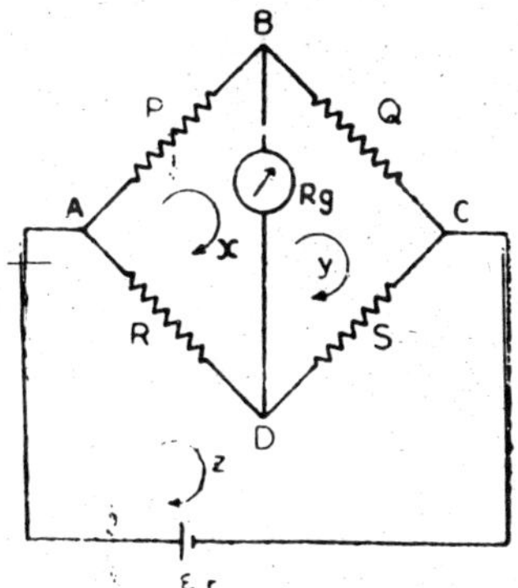

Fig. 2.17

Collecting the coefficient of x, y and z in equation (1) and (2), we have,

$$x(P + R + R_g) - yR_g - zR = 0$$

and $$-xR_g + y\,(Q + S + R_g) - zS = 0$$

Multiplying the first of these equations by S and second by R and subtracting, we get,

$$x[S(P + R + R_g) + RR_g] - y[SR_g + R(Q + S + R_g)] = 0$$

The current in the galvanometer = x − y. At the balance point this is zero, *i.e.*, $x - y = 0$

or $\quad x = y.$

Hence, we get,

$$S(P + R + R_g) + RR_g = SR_g + R(Q + S + R_g)$$

or $\quad SP = RQ$

hence, $\dfrac{P}{Q} = \dfrac{R}{S}$

which is the required condition for balanced Wheatstone's bridge given in equation (2.39).

2. Kelvin's Method for Determining the Resistance of a Galvanometer

In this method the galvanometer is put in the unknown arm of the post office box which is a Wheatstone bridge. The points B and D are connected by a tapping key K. Let the clockwise cyclic currents through the meshes, (1) ABDA, (2), BCDB and (3) ABCEA be x, y and z respectively as shown in Fig. 2.18.

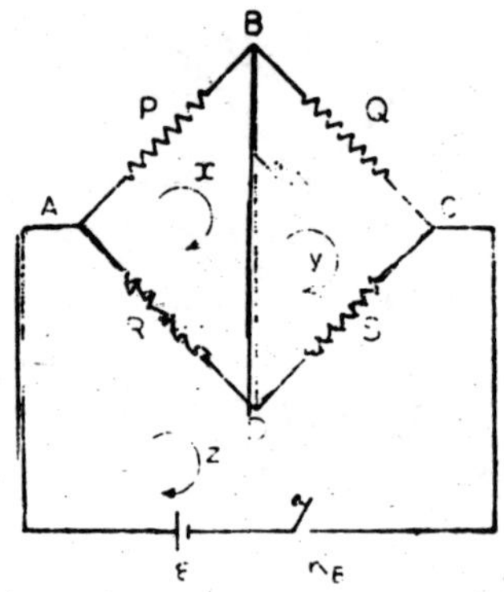

Fig. 2.18

Let P, Q, R, S, G and B be the resistance of the arms AB, BC, AD, DC BD and AC respectively. Applying Kirchhoff's second law to the meshes ABDA and BCDA, we get,

$$xP + (x - y) G + (x - z) R = 0 \quad ...(1)$$

$$yQ + (y - z) S + (y - x) G = 0 \quad ...(2)$$

The values of resistance P, Q and R are adjusted such that there is no charge in the deflection of the galvanometer when the key K_B is opened or closed.

This means that current in this arm must be zero *i.e.,*

$$x - y = 0$$

or $$x = y,$$

Substituting in equation (1) and (2), we get,

$$yP + (x - z) R = 0$$

$$yQ + (y - z) S = 0$$

or $$x(P + R) = zS$$

and $$y (Q + S) = zR$$

or ratio, $$\frac{x(P+R)}{y(Q+S)} = \frac{R}{S}$$

or $$\frac{P+R}{Q+S} = \frac{R}{S} \text{ since } x = y$$

$$PS + RS = RQ + RS$$

or $$S = R\frac{Q}{P} \quad ...(2.42)$$

Thus from the known values P, Q and R, S can be calculated.

Adjustment

(1) When the battery key K_B is pressed the galvanometer will always show a deflection as it is in one of the arms of the bridge. Generally, the deflection is out of scale, it may be reduced and brought within scale by connecting a high resistance box in the battery circuit.

(2) When the tapping key K_B is pressed the galvanometer deflection will change as some of the current will also flow in BD arm. P/Q is a fixed ratio for one observation, hence R is adjusted such that the galvanometer deflection remains the same before and after pressing key K_B. *In fact this method is a constant deflection*

method and not to be confused with null method. This method was first employed by Kelvin hence it is called Kelvin's method.

MANCE'S METHOD OF MEASURING THE INTERNAL RESISTANCE OF A CELL

Wheatstone's bridge circuit has been employed by Mance's to determine the internal resistance of the cell as shown in Fig. 2.19. The cell is connected in the unknown arm of the bridge and usual arm of the cell is connected by a key K_B. Let P, Q, R, r and G be the resistance of the arms AB, BC, DA, CD, and BD respectively.

Applying Kirchhoff's second law to the meshes ABDA and BCDB, we get,

$$(i - i_g - i_1) P - i_g G + (i - i_g) R = 0 \qquad ...(1)$$

$$ir + i_g G + (i - i_1) Q = x \qquad ...(2)$$

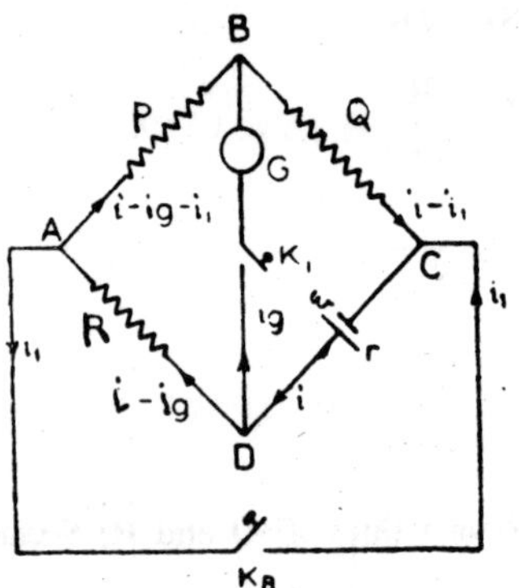

Fig. 2.19

Collecting the coefficients of i, ig and i_1, we get,

$$i(P + R) - i_g(P + G + R) - i_1 P = 0 \qquad ...(3)$$

$$i(r + Q) + i_g G - i_1 Q = x \qquad ...(4)$$

Multiplying (3) by (r + Q) and (5) by (P + R), we get,

$$i(P + R)(r + Q) - i_g(P + G + R)(r + Q) - i_1 P(r + Q) = 0 \qquad ...(5)$$

$$i(r + Q)(P + R) + i_g G(P + R) - i_1 Q(P + R) = x(P + R) \qquad ...(6)$$

subtracting (5) from (6),

$$i_g[G(P+R) + (P+R)(r+Q)] - i_1\{Q(P+R) - P(r+Q)\} = x(P+R) \qquad ...(7)$$

Adjustment

The values of P, Q, and R are adjusted as in the case of Kelvin's method (the procedure is exactly the same) such that the galvanometer shows a "constant deflection" when the key K_1 is pressed or opened. That is the current in the galvanometer is constant, The equation (7) shows that ig will remain constant whatever be the value of current i1 if the coefficient of i_1 is zero.

$$Q(P + R) - P(r + Q) = 0$$

or $$Q(P + R) = P(r + Q)$$

$$PQ + QR = Pr + PQ$$

$$r = \frac{Q}{P} R \qquad ...(2.43)$$

Hence r the internal resistance of the cell can be calculated by knowing P, Q and R.

KELVIN'S DOUBLE BRIDGE FOR LOW RESISTANCE

The Wheatstone's bridge is not suitable for comparing two very low resistance such as metal rods because the junction resistances and resistances of connecting wires are larger compared to the low resistance to be measured. For accurate determination of low resistance, Kelvin devised double bridge as shown in Fig. 2.20.

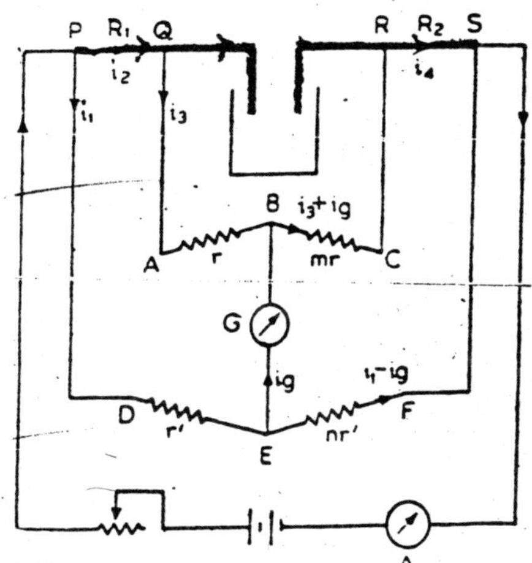

Fig. 2.20

PQ and RS are two low resistances of the same order of magnitude such as the *thick conducting rods of uniform cross-section.* One of which may be the known variable standard say PQ resistance and the other (RS) is unknown resistance. The rods PQ and RS are connected in series through a mercury cup. The bridge is completed by two resistance coils AC and DF whose middle points are at B and E which are connected to the rods PQ and RS as shown in Fig. 2.20.

Let the resistance coil ABC be divided at B in the ratio r : mr and the coil DEF in the ratio r': nr'. Let the currents in various conductors be those indicated in the figure. If i' be the current between Q and R then at the junction Q the Kirchhoff's current law gives,

$$i_2 - i_3 - i' = 0$$

or $$i' = i_2 - i_3 \qquad ...(1)$$

and at the junction R, we get,

$$i' + (i_3 + i_g) - i_4 = 0$$

or $$i' = i_4 - (i_3 + i_g) \qquad ...(2)$$

Equating (1) and (2),

$$i_2 - i_3 = i_4 - (i_3 + i_g)$$

or $$i_g = i_4 - i_2 \qquad ...(3)$$

Applying Kirchhoff's law to the meshes PQABEDP and RSFEBCR, we get,

$$i_2R_1 + i_3r - i_gG - i_1r' = 0 \quad ...(4)$$

$$i_4R_2 - (i_1 - i_g)\, nr' + i_gG + (i_3 + i_g)\, mr = 0 \qquad ...(5)$$

The point P of the variable resistance which is in the form of knife edge so that it can slide along the rod is so adjusted that there is no deflection in the galvanometer *i.e.,* when $i_g = 0$, then equation (4) and (5) become,

$$i_2R_1 + i_3r - i_1r' = 0 \qquad ...(6)$$

$$i_4R_2 - i_1nr' + i_3nr = 0 \qquad ...(7)$$

But from equation (3), when ig = 0

$$i_4 = i_2$$

therefore,

$$i_2R_1 + i_3r - i_2r' = 0$$

$$i_2R_2 - i_1nr' + i_3mr = 0$$

or $$i_2R_1 = i_2r' - i_3r$$

$$i_2R_2 = i_1nr' - i_3mr$$

or $$\frac{R_2}{R_1} = \frac{i_1nr' - i_3nr}{i_1r' - i_3r} \quad ...(2.44)$$

If the points B and E divide the two coils ABC and DEF in the same ratio, *i.e.,* when m = n, then,

$$\frac{R_2}{R_1} = m = n \quad ...(2.45)$$

Usually, we make,

$$m = n = 1$$

then, $R_2 = R_1$.

Thus, by suitably adjusting the variable point P of the known resistance, R_1 for zero deflection in the galvanometer, we get directly the unknown resistance R_2. This bridge is suitable to measure resistances from 0.1 to 0.003 ohm.

POTENTIOMETER

A potentiometer is a device used to measure potential difference. The device is simply to compare an unknown emf with a known one. It consists of a uniform wire AB of resistance R. A steady current is passed through the wire AB by a constant source of emf ξ through the rheostat Rh. ξ_1 is emf of the cell which is to be measured and is connected in the circuit as shown in Fig. 2.21. When the current is steady and constant, the potential gradient is constant.

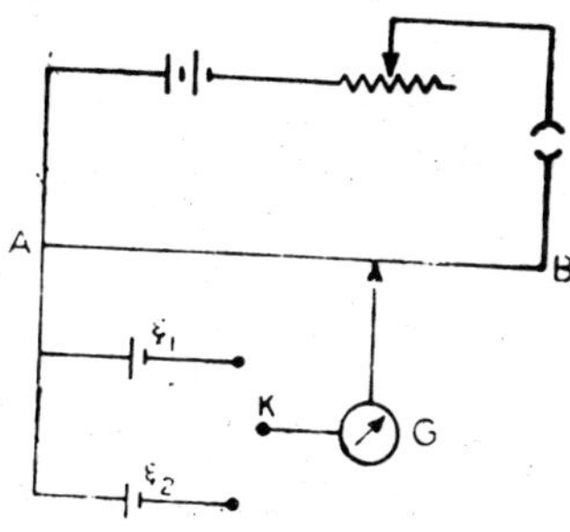

Fig. 2.21

Let resistance per unit length of the wire AB be ρ and the steady current passing through the wire be I. Then the potential gradient that is the fall of potential per unit length of the wire is I ρ. Connect ξ_1 by K and move the sliding contact till the galvanometer gives no deflection. If l_1 is the length of the potentiometer wire for balancing ξ_1, then

$$\xi_1 = I\rho l_1.$$

Similarly connect x^2 by K and repeat the process, now if l_2 is the length of the potentiometer wire for balancing x_2, then,

$$\xi_2 = I\rho l_2.$$

Hence ratio,

$$\frac{\xi_1}{\xi_2} = \frac{I\rho l_1}{I\rho l_2} = \frac{l_1}{l_2}. \qquad ...(2.46)$$

If one of the cells say ξ_2 is known (a standard cell) the emf of the other can be known. The equation (2.46) shows that for a steady current passing through the potentiometer wire, the potential difference across any length is proportional to the length of the wire.

Sensitiveness of a Potentiometer

In potentiometer, *the accuracy increases as the length of the wire is increased. But a potentiometer of greater length becomes unwieldy.* This difficulty is overcome by using compact forms of potentiometer.

TYPES OF POTENTIOMETERS

Many different types of compact forms have been designed. The principle of working of all types of potentiometers is the same.

The commonly used potentiometers are : (i) Crompton (ii) Student and (iii) Rayleigh potentiometer.

Crompton Potentiometer

Since it is difficult to get a very long, uniform homogenous wire, hence it is divided equally into fifteen segments. One of the wires is stretched on a wooden board and rest of the fourteen segments are repeated by the coils connected in series with the stretched wire, the resistance of each of coil is equal to the wire as shown in Fig. 2.22.

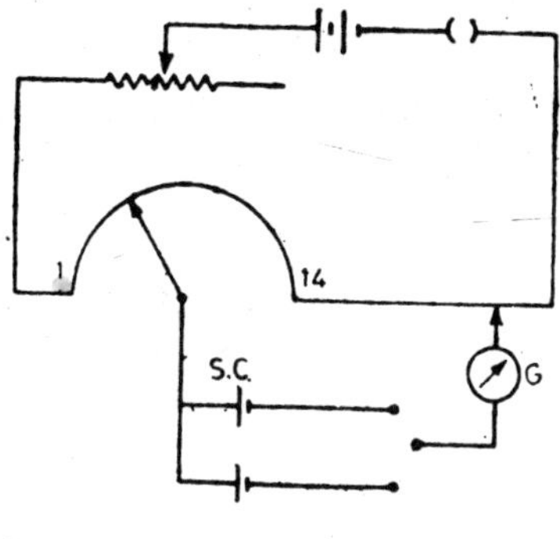

Fig. 2.22

Student Potentiometer

This type of potentiometer was designed by Dr. N.K. Sethi. In this potentiometer, a platinum wire of length 50 cm and of uniform cross-section is used. Its resistance is 1 ohm. There are ten coils each of one ohm resistance put in series with the wire and on the scale each cm. is indicated as two.

Standardisation of Potentiometer of Direct Reading Potentiometer

A standard cell (cadmium cell) is connected as shown in Fig. 2.23 the resistance is adjusted to 1018.3 ohm and the variable resistance (rheostat) so adjusted as to give no deflection with the standard cadmium cell in circuit. The current will be given by,

$$\frac{1.0183}{1018.3} = \times\ 10^{-3} \text{ amp,}$$

where 1.0183 is the emf of the cadmium cell in volt. Now if the resistance with AB = 1 ohm as in the case of student potentiometer, then the potential difference per mm of the scale,

$$= 10^{-3} \times 0.001 \text{ volt/mm}$$

$$= 1 \text{ micro volt/mm} = 1 \text{ mili. volt/cm}$$

When the wire is calibrated to read one millivolt per cm the potentiometer will read the current in milliamperes per cm. *Thus, we can read current and voltage directly on the scale and such device is also called " direct reading" Potentiometer.*

APPLICATIONS OF POTENTIOMETER

Some important uses of potentiometer are discussed below:

1. Calibration of an Ammeter and Voltmeter

The principle of standardisation is employed for the calibration of an ammeter and voltmeter. Connections are made as shown in Fig. 2.23.

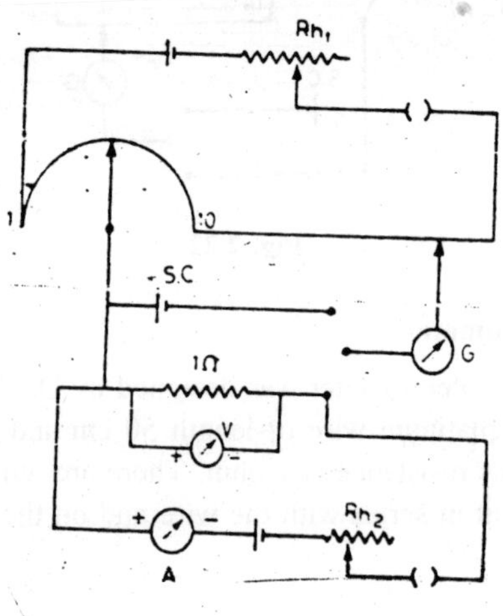

Fig. 2.23

First the potentiometer wire is calibrated in milli-volt by obtaining the balance point with the standard cell of emf 1.0183 volts at 1018.3 cm. The balance point is obtained by varying Rh_1. Next with the help of Rh_2, the current in the ammeter A is adjusted and corresponding to this value of current the balance point on the potentiometer is noted. Then the current is calculated *i.e.,* directly the reading is current in amp. when multiplied by 10^{-3}. Similarly the voltmeter gives the voltage. The experiment is repeated for various readings of the ammeter and voltmeter. A calibration graph between the observed and calculated values is drawn.

2. Measurement of Very Small Thermo emf

The emf of the order of 30 or 40 milli-volts e.q. thermo emf can be measured with potentiometer. First the potentiometer wire is calibrated in milli-volts with standard cell as discussed above. The connection are made as shown in Fig. 2.24.

Let us take copper-iron couple to produce the thermo emf. As the potentiometer wire is calibrated *i.e.,* the p.d. per mm of the scale =

$10^{-3} \times 0.001$ volt = 1 microvolt/mm = 1 milli-volt/cm. We can read directly the thermo-emf generated at various temperatures.

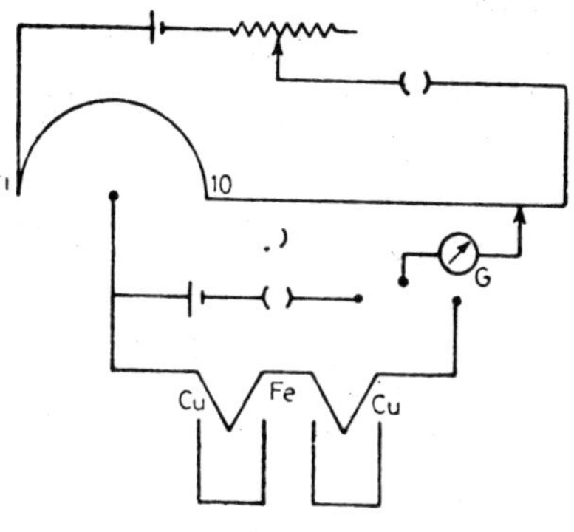

Fig. 2.24

3. Measurement of Large Potential Difference

Though, the potentiometer is best suited for measuring the potential difference of the order of less than 2 volts, we can measure large potential difference by dividing it into fractions by volt box. A volt box consists of a very large resistance, about 10,000 ohms. This resistance is divided in fractions by intermediate tappings. The large potential to be measured is applied to this resistance shown in Fig. 2.25.

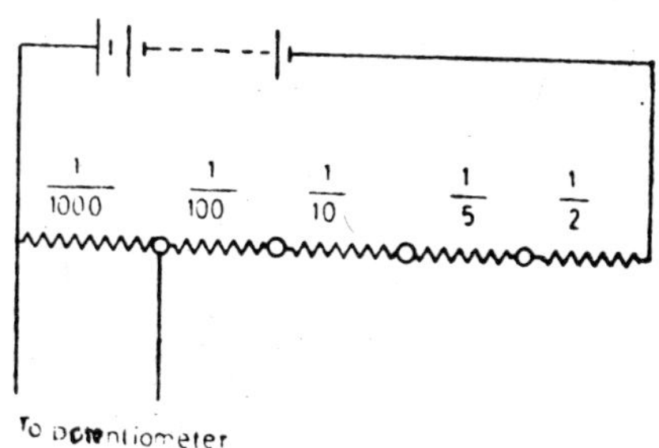

Fig. 2.25

Take any fraction, say 1/100 and balance it against the potentiometer wire and the p.d. for this fraction is known, then multiply it by 100 to obtain the requisite high potential.

4. Determination of Internal Resistance of a Cell

The connections are shown in Fig. 2.26.

First the key K is open and the emf, ξ of the cell is balanced, say l_1 is the balancing length of the potentiometer wire. Now close the key K and introduce the resistance in the box R and the potential drop V = IR is balanced against a length l_2 of the potentiometer wire. If r is the internal resistance of the cell, then,

$$\frac{\xi}{V} = \frac{I(R+r)}{IR}$$

$$= \frac{R+r}{r} = \frac{l_1}{l_2}$$

or $$r = \frac{l_1 - l_2}{l_2} R. \quad ...(2.47)$$

Hence r can be calculated.

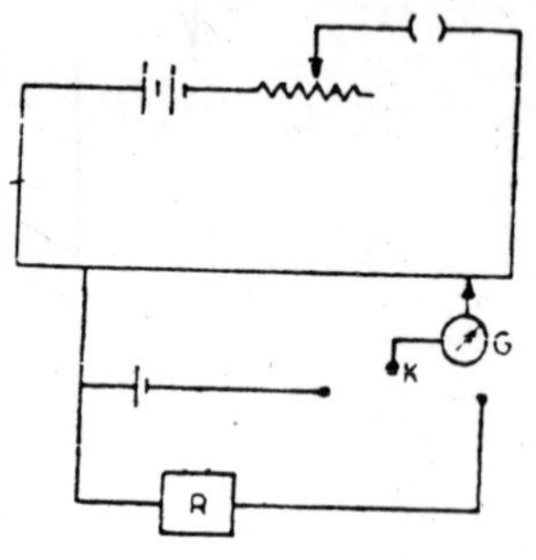

Fig. 2.26

5. Comparison of Two Low Resistances

Two low resistances of the same order can be compared using a potentiometer. The connections are made as shown in Fig. 2.27. As the same current is flowing through R_1 and R_2, then potential drops are IR_1 and IR_2 respectively. If the balancing lengths on the potentiometer wire are l_1 and l_2 respectively, then,

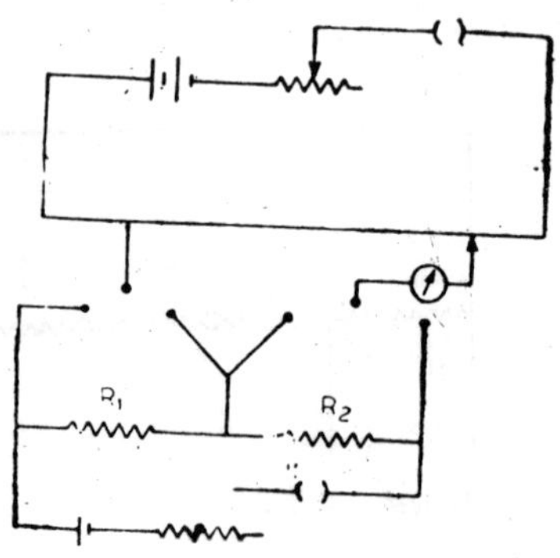

Fig. 2.27

$$\frac{IR_1}{IR_2} = \frac{R_1}{R_2} = \frac{l_1}{l_2} \quad ...(2.48)$$

6. Determination of a Given Low Resistance

If in the above experiment one of the resistance (R_2), is standard resistance then the other (R_1) can be calculated by relation

$$R_1 = R_2 \frac{l_1}{l_2}. \qquad ...(2.49)$$

SOLVED EXAMPLES

Example 1:

A battery of 6 volts emf and 0.5 ohm internal resistance is joined in parallel with another of 10 volts emf and 1 ohm internal resistance. The combination is used to send a current through an external resistance. The combination is used to send a current through an external resistance of 12 ohm. Calculate, by application of Kirchhoff's law, the current through each battery.

Solution:

Let the current through the batteries B_1 and B_2 be I_1 and I_2 repulsively [Fig. 2.28]. Then applying Kirchhoff's first law to the point C, the current through the external resistance is $(I_1 + I_2)$.

Applying Kirchhoff's second law to the mesh ABFEA

$$I_1 \times 0.5 + (I_1 + I_2)\, 12 = 6$$

or $$12.5\, I_1 + 12\, I_2 = 6$$

$$25\, I_1 + 24\, I_2 = 12 \quad ...(1)$$

Similarly applying Kirchhoff's second law to the mesh CDFEC,

$$I_2 \times 1 + (I_1 + I_2)\, 12 = 10$$

or $$12\, I_1 + 13\, I_2 = 10 \quad ...(2)$$

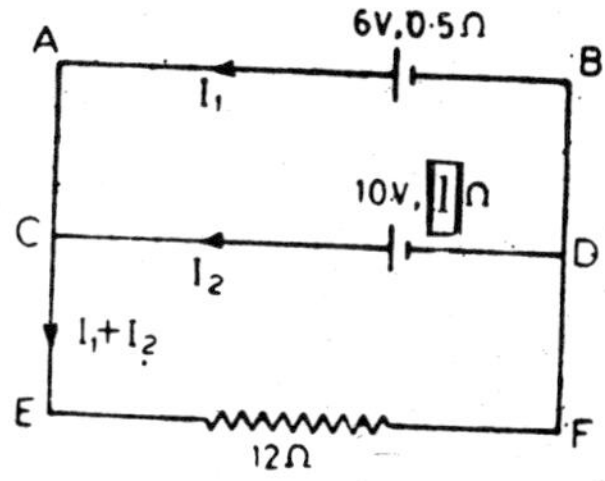

Fig. 2.28

Solving (1) and (2), we get,

$$I_1 = -2.27 \text{ amp}; \; I_2 = 2.865 \text{ amp}.$$

The negative sign in I_1 indicates that the current I_1 in the battery B is opposite to the direction shown in Fig. 2.16.

Example 2:

12 wires, each of resistance r ohms, are connected in the form of a skeleton cube. Find the equivalent resistance of the cube when the current enters at one corner and leaves at the diagonally opposite corner.

Solution:

Let a current of 6I enter the pint A and leave the point G at the opposite end of the diagonal.

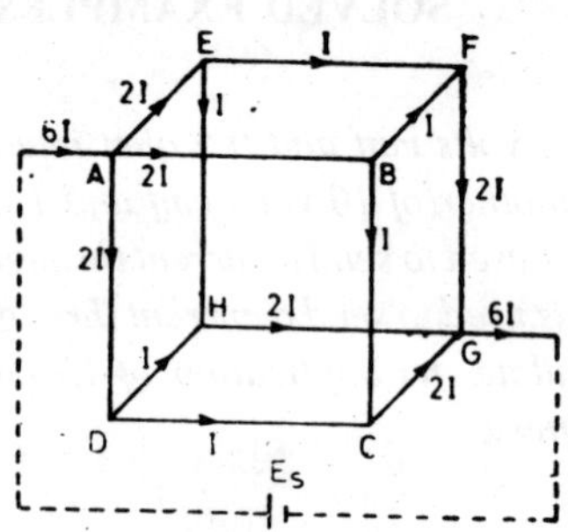

Fig. 2.29

The distribution of the current is shown in Fig. 2.29 as well as equivalent circuit and let R be the equivalent resistance.

Applying Kirchhoff's law to any path between AG, say ABFG, we get,

$$2Ir + Ir + 2Ir = V = 6IR$$

or $$5Ir = 6IR$$

or $$R = \frac{5}{6}\ r.$$

Example 3:

12 wires, each of resistance r ohms, are connected in the form of a skeleton cube. Find the equivalent resistance of the cube when a cell is joined across any one of the wires forming the cube.

Solution:

Applying Kirchhoff's law to the mesh FCGHD and ABCDA as shown in Fig. 2.30,

$$zr - (y - z)\ r - 2\ (y - z)\ r - (y - z)\ r = 0$$

and $$xr - yr - zr - yr = 0$$

or $\quad 5z - 4y = 0$

or $\quad z = \dfrac{4}{5}\,y$

and $\quad x - 2y - z = 0$

or $\quad x - 2y - \dfrac{4}{5}\,y = 0$

i.e., $\quad y = \dfrac{5}{14}\,x.$

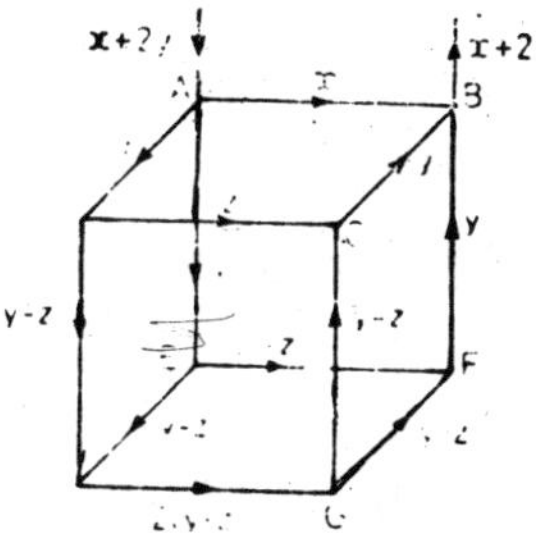

Let the equivalent resistance be R.

Fig. 2.30

Then,

P.D. across AB = $(x + 2y)\,R = xr.$

Therefore,

$$R = \frac{x}{x+2y}\,r = \frac{x}{x+\frac{10}{14}x}\,r = \frac{7}{12}\,r$$

Example 4:

What is the equivalent resistance between the terminal points x and y in the network shown in Fig. 2.19? Assume that the resistance of each resistor is 10 ohms.

Solution:

Assume the current distribution as shown in Fig. 2.31. Applying Kirchhoff's first law at C,

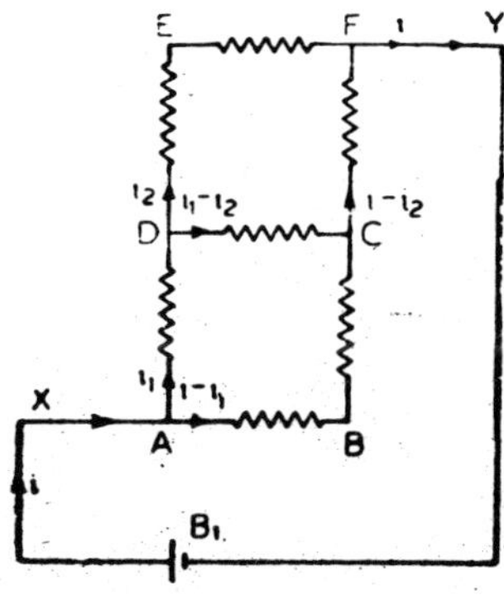

Fig. 2.31

i.e., $(i - i_1) + (i_1 - i_2) - (i - i_2) = 0$

or $i - i_2 = (i - i_1 + i_1 - i_2)$.

This is the current along CF.

Now again applied the Kirchhoff's second law to the closed loops, ADCBA and CDEFC, we get,

$$i_1R + (i_1 - i_2) R - (i - i_1) 2R = 0$$

or $4i_1 - i_2 = 2i$ ...(1)

and $2i_2R - (i - i_2) R - (i_1 - i_2) R = 0$

or $-i_1 + 4i_2 = i$ ...(2)

Solving (1) and (2), we get

$$i_1 = \frac{3}{5} i \text{ and } i_2 = \frac{2}{5} i \qquad ...(3)$$

Again applying second law to mesh $XABCFYB_1X$

$$(i - i_1) 2R + (i - i_2) R = x.$$

Substituting i_1 and i_2 from (3), we get,

$$\left(i - \frac{3}{5} i\right) 2R + \left(i - \frac{2}{5} i\right) R = x$$

or $\frac{7}{5} iR = x$

If r is the equivalent resistance between the terminal points x and y, then,

$$x = ir$$

Hence, $r = \frac{7}{5} R$.

As given R = 10, hence

$$r = \frac{7}{5} \times 10 = 14 \text{ ohms.}$$

Example 5:

A source of emf ξ volts and internal resistance r ohms, supplies current to a heating coil. Calculate the resistance R of the heating coil in terms of r so that,

(1) *heating takes place most rapidly,*

(2) *3/4 of the total energy delivered by the source goes into the heating coil.*

Solution:

The current through the heating coil

$$i = \frac{\xi}{R+r} \qquad ...(1)$$

The rate of energy delivered by the source to the heating coil,

$$W = i^2R = \left(\frac{\xi}{R+r}\right) R = \frac{\xi^2 R}{(R+r)^2} \qquad ...(2)$$

(1) As we have discussed in equation (2.16), for maximum power, the condition is that R = r.

(2) The rate of total energy delivered by the source,

$$= x_i = \frac{\xi^2}{R+r}$$

As given, $\dfrac{\xi^2 R}{(R+r)^2} = \dfrac{3}{4}\dfrac{\xi^2}{R+r}$

hence, after simplification, we get,

R = 3r.

Example 6:

A silver wire 1 mm in diameter carries a charge of 90 coulombs in 1 hour and 15 minutes. Silver contains 5.8 × 10^{22} free electrons/cm^2. Calculate, (i) the current in the wire in amperes, (ii) the drift velocity of the electrons in wire in metres/sec.

Solution:

(1) 1 hour and 15 minutes = 4500 secs.

Current i = q/t

i = 90/4500 = 0.02 ampere

(2) Current density $\vec{j} = \dfrac{\text{Current}}{\text{Area}}$

$$= \frac{0.02\text{amp}}{\pi \times (0.0005)^2 \text{ m}^2}$$

$$= \frac{0.02}{3.14 \times 5 \times 5 \times 10^{-8}} \text{ amp/m}^2$$

= 2.55 × 10^4 amp/m^2

The drift velocity,

$$v_d = \frac{J}{ne} = \frac{2.55\times10^4}{1.6\times10^{-19}\times5.8\times10^{28}} \text{ m/sec.}$$

(Since 1 cm^3 = 10^{-6} m^3)

hence, n = 5.8 × 10^{28} elect.m^3

v_d = 2.69 × 10^{-6} metre/sec.

Example 7:

In a neon gas discharge tube 2.9 × 10^{16} Ne + ions move to the right through a cross-section of the tube each second, while 1.2 × 10^{18} electrons move to the left in the same time. Find the magnitude and the direction of the current.

Solution:

The charge on each neon ion, q_1 = e 1.6 × 10^{-19} coul.

Let us take the positive direction from left to right and negative from right to left.

The current due to Ne + ions = $\frac{n_1 q_1}{t_1}$

$$= \frac{2.9\times10^{18}\times1.6\times10^{-19}}{1} \text{ coul/sec.}$$

= 0.446 amp.

The charge on each electron, q_2 = – 1.6 × 10^{-19} coul.

The current due to electrons

$$= -\frac{n_2 q_2}{t_2} = -\frac{(1.2\times10^{18})(-1.6\times10^{-19})}{1}$$

= 1.2 × 1.6 coul/sec = 1.92 amp.

Therefore, the net current through the discharge tube

= 0.446 + 1.92 = 2.366 amp.

Thus, the magnitude of the current is 2.366 amp. and its direction is from left to right.

Example 8:

An aluminium wire whose diameter is 0.245 cm. is welded end to end to a copper wire diameter 0.1626 cm. The composite wire carries a steady current of 10 amp.

(a) *What is the current density in each wire?*

(b) *Find the drift speed of the electrons in copper wire. Assume one free electron per atom in copper.*

Given Avogadro Number = 6×10^{23} *atoms per gm-mole*

Molecular weight of copper = *64*

Density of copper = *9 gm/cc.*

Solution:

(a) The current density is given by

$$\vec{J} = \frac{i}{A}.$$

The area of aluminium wire, $A_1 = \pi r_1^2$

$$= 3.14 \left[\frac{0.245 \times 10^{-2}}{2}\right] = 4.713 \times 10^{-6} \text{ m}^2$$

The area of copper wire, $A_2 = \pi r_2 2$

$$= 3.14 \left[\frac{0.1626 \times 10^{-2}}{2}\right]^2 = 2.075 \times 10^{-6} \text{ m}^2$$

Therefore, the current density in aluminium wire,

$$J_{Al} = \frac{i}{A_1} = \frac{10}{4.713 \times 10^{-6}} = 2.12 \times 10^6 \text{ amp/m}^2$$

similarly, the current density in copper wire,

$$J_{cu} = \frac{i}{A_2} = \frac{10}{2.075 \times 10^{-6}} = 4.818 \times 10^{-6} \text{ amp/m}^2$$

(b) The number of atoms per unit volume = rN/M where r is the density of copper, N is Avogadro number and M is atomic weight of copper.

A given, there is one free electron per atom in copper, the number of free electrons per unit volume,

$$n = \frac{\rho N}{M}$$

$$= \frac{9 \times (6 \times 10^{23})}{6} \text{ electrons/cm}^3$$

$$= 8.437 \times 10^{28} \text{ electrons/m}^3$$

But, $J = nev_d$

Therefore

$$v_d = \frac{J_{cu}}{n_e} = \frac{4.818\times 10^6}{8.437\times 10^{23}\times 1.6\times 10^{-19}}$$

$= 2.834 \times 10^{-4}$ m/sec.

Example 9:

The potential difference across a wire of 10^{-3} cm^2 cross-sectional area and 50 cm length is 2 volt, when a current of 0.25 amp. exists in the wire. Calculate,

(a) *The field strength in the wire,*

(b) *The current density, and*

(c) *The conductivity of the metal*

Solution:

(a) The field strength in the wire is given by,

$$E = \frac{V}{d} = \frac{2\text{volt}}{50\text{cm.}} = \frac{2\text{volt}}{0.5\text{metre}} = 4 \text{ volts/metre.}$$

(b) The current density,

$$J = \frac{i}{A} = \frac{0.25\text{amp}}{10^{-3}\,\text{cm}^2}$$

$$= \frac{0.25\text{amp.}}{10^{-3}\times 10^{-4}\,\text{m}^2}$$

$= 2.5 \times 10^6$ amp/m².

(c) The electrical conductivity of metal s is given by

$J = \sigma E$

or $$\sigma = \frac{J}{E} = \frac{(2.5\times 10^6\,\text{amp./m}^2)}{(4\,\text{volts / metre})}$$

$= 6.25 \times 10^5$ mho/metre.

Example 10:

Calculate the insulation resistance of a cable having a cylindrical conductor of radius r_1 and the outer radius of insulation r_2. The specific resistance of the insulation is ρ.

Solution:

Let r and r + dr be the radii of thin concentric layer of insulation, then the surface area = $2\pi rl$, where l is the length of the cable and resistance of the layer = $\frac{\rho dr}{2\pi rl}$, according to eq. (2.30).

Therefore, total resistance of the cable of length l

$$R = \int_{r_1}^{r_2} \frac{\rho dr}{2\pi rl} = \frac{\rho}{2\pi l} \log_e \frac{r_2}{r_1}$$

Example 11:

Find the resistance of 1 cc of copper when drawn into a wire of diameter 0.32 mm specific resistance of copper = 1.59 × 10^{-6} ohms. cm

Solution:

Radius r of the wire = 0.016 cm

Length of the wire = $\frac{\text{Volume}}{\pi r^2} = \frac{1}{3.14 \times 0.016^2}$

and according to equation (2.30),

$$R = \frac{\rho l}{a} = \frac{1.59 \times 10^{-6}}{3.14 \times 0.016^2 \times 3.14 \times 0.016^2}$$

= 2.46 ohm.

EXERCISES

1. A current of 1 amp is passing through a copper wire 10 metre in length and 0.08 mm in diameter. If the resistivity of copper is 1.65 × 10^{-8} ohm-metre, calculate

   (1) The resistance of copper wire; and

   (2) The potential difference between the ends of the wire.

2. In copper there are 10^{23} free electrons per cc. all of which contribute to a current of 1 amp in a wire of copper of 0.01 cm^2 cross-sectional area.

   (1) What is the average drift speed of electrons in the copper?

   (2) What is the electric field in the wire?

   Given specific resistance of copper = 1.6 × 10^{-8} ohm. metre.

3. In a cloud of electrons having 6×10^{29} electrons per metre^3, 40 per cent of the electrons have a drift velocity 1 mm per second in x-direction, 40 per cent have drift velocity of 2 mm per second in y-direction and remaining 20 per cent have drift velocity of 3 mm per second in z-direction. Calculate the current density $\vec{J}$.

4. (a) What is the electric field strength in a copper conductor of cross-sectional area 1 cm2 when a current of 200 amp. flows through it? The resistivity of copper is 1.7×10^{-8} ohm-metre.

   (b) Also calculate the potential difference between point 2 metres apart along the length of the above conductor.

5. An aluminium wire of diameter 0.10 cm is welded end to ent to a copper wire of diameter 0.064 cm. The composite wire carries a steady current of 10 amp. What is the current density in each wire? What is the drift velocity of electrons in copper density in each wire? Given that the density of copper is 9.0 gms/se and Avogadro's number is 6×10^{23} atoms/mol.

6. (a) Let n resistances R_1, R_2 ... R_n be connected in series. Show that the equivalent resistance R is given by,

$$R = \sum_{i=1}^{n} R_i$$

   (b) Let n resistances R_1, R_2 ... R_n be connected in parallel. Show that the equivalent resistance R is given by,

$$\frac{1}{R} = \sum_{i=1}^{n} \frac{1}{R_i}.$$

7. Define emf and potential difference. What is the difference between an emf and potential difference?

8. What do you mean by electric current and current density? How are they related to each other? Establish the relation between current density and the velocity of charge carrier.

9. What is the conservation of charge rule? Show that

$$\nabla.\vec{J} + \frac{\delta p}{\delta t} = 0.$$

10. Define (a) resistance, (b) resistivity, (c) conductivity. Establish the relation $\vec{J} = \sigma\vec{E}$.

11. What is the relation between conductivity and permittivity? Show that R = $\varepsilon/\sigma c$.
12. State and prove Kirchhoff's laws of distribution of currents in an electrical network and apply these laws to deduce the condition of
    (a) Balance of Wheatstone's bridge.
    (b) Maximum sensitivity of Wheatstone Bridge.
13. Explain the Maxwell's cyclic currents method. Apply it to obtain the Wheatstone's bridge condition.
14. Give Kelvin's method for measuring the galvanometer resistance.
15. Describe how you will measure the internal resistance of a cell using Wheatstone's bridge.
16. Why is Wheatstone's bridge unsuitable for measurement of very low resistance? Describe Kelvin's double Bridge method for determining a low resistance.
11. Give the principle of a potentiometer.
17. Describe the construction of a Crompton potentiometer. How is it used for measuring
    (1) Large potential difference
    (2) Very small thermo emf
    (3) Low resistance.

Multiple Choice Questions

1. The ampere represents a specific amount of electrons passing given point in a circuit in a specific.
   (a) direction (b) quantity
   (c) time (d) quantity and time
2. The sensitivity of a Wheatstone bridge is maximum when
   (a) P/Q = R/X (b) R = X, = P = Q
   (c) P=Q=R=X=G (d) None of these
3. If you were to double the length of wire, the resistance would
   (a) half (b) remain the same
   (c) resistivity (d) permittivity

4. Electrical conductivity is the reciprocal of

   (a) conductance (b) remance

   (c) resistivity (d) permittivity

5. The unit of conductance is

   (a) Henry (b) Ohm

   (c) Mho (d) Farad

6. The reciprocal of the resistance is termed

   (a) Admittance (b) Suscepectance

   (c) Conductance (d) Conductivity

7. A potentiometer works is the principle that an unknown ... is measured by balancing it, wholly or in part against a known.

   (a) e.m.f. (b) current

   (c) resistance (d) reactance

8. While using a Wheatstone bridge

   (a) both the keys should be pressed simultaneously

   (b) battery key should be pressed first

   (c) galvanometer key should be pressed first

   (d) none of these

9. The low resistance can be most accurately measured by

   (a) The Wheatstone bridge method

   (b) Kelvin's bridge method

   (c) Potentiometer

   (d) None of the above

10. For steady state continuity equation is

    (a) $\vec{\nabla}.\vec{J} = 0$ (b) $\tilde{N}\ \vec{J} = -\frac{d\rho}{dt}$

    (c) $\tilde{N}\ \vec{J} = 0$ (d) $\nabla.\vec{J} = \frac{d\rho}{dt}$

11. Kirchhoff's laws are applicable to

    (a) D.C. only

    (b) A.C. only

(c) A.C. and D.C. both

(d) None of the above

12. Closed circuit technique are based on

(a) Thevenin theorem

(b) Norton theorem

(c) Kirchhoff's current law

(d) Kirchhoff's voltage law.

13. In cells

(a) Chemical energy changes into mechanical

(b) Chemical energy changes into heat

(c) Chemical energy changes into electricity

(d) None of the above.

14. E.M.F. of a cell is always greater than P.D.

(a) True (b) False

15. If there is a regular movement of ... within a material, we say that an electric current if flowing

(a) atoms (b) ions

(c) electrons (d) protons

16. $\vec{J} = \sigma\vec{E}$ is

(a) A Maxwell's equation

(b) continuity equation

(c) Ohm's law

(d) Ampere's law

17. Ohm's law does not apply to

(a) a.c. circuits

(b) conductors

(c) semi-conductors

(d) Conductors when there is a change in temperature

3

Electrometers and Electrostatics Machines

ELECTROMETERS

In common practice we use the *electromagnetic voltmeters* for measuring the potential difference, but they are *less accurate* because they *require some current,* though very small, *for their working.* In electrostatics phenomena, certain instruments have been devised, in which the *current is to be completely avoided* by perfect insulation which is difficult to achieve. However, the efforts have been made to reduce, the leakage of charge due to imperfect insulation, to minimum. Thus, *accuracy is increased and we measure correctly the potential difference,* charge and capacitance. *Such devices are called electrometers.* The simplest electrometer was first used by Faraday for measuring potential difference and that was gold leaf electroscope.

KELVIN'S ABSOLUTE OR ATTRACTED DISC ELECTROMETER

Principle : The principle of attraction between the plates of charged capacitor was utilized by Kelvin for constructing the attracted disc-electrometer.

Construction : It is essentially a guard ring air condenser consisting of two metallic plate A and B and guard ring G to avoid edge effect as shown in Fig. 3.1. The central plate A which can freely pass concentrically through G serves as the attracted disc and is held in position with the help of spring S. Spring S is connected to a rod which can be moved up or down with micrometer screw N. Hence plate A can be moved by screw N.

The plate A and guard ring G are always charged to the same potential and the plate B is either earthed or charged to a different potential. In order to obtain uniform electric field between A and B, the plate A is connected to g by light flexible wire, then the lines of force will be parallel and equidistant.

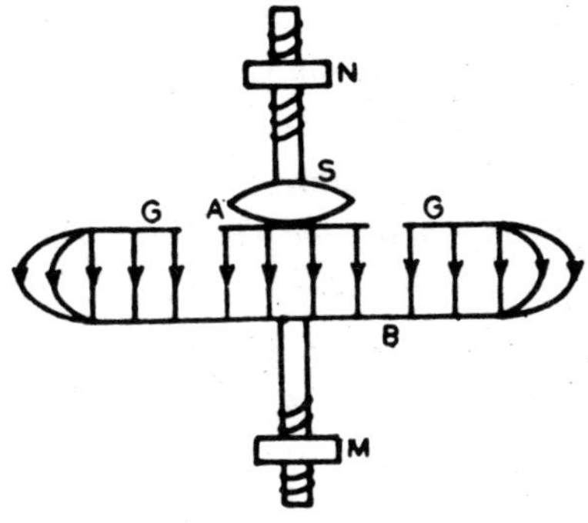

Fig. 3.1

Theory : Let V_a and V_b be the potentials of plates A and B respectively and d be the distance between them, then the electric field,

$$E = \frac{V_a - V_b}{d} \quad ...(1)$$

Now the plate A will experience a force of attraction per unit area *i.e.,* pressure given by

$$P = \frac{\varepsilon_o E^2}{2}$$

Therefore, the plate A will experience a force of attraction.

$$F = \frac{\varepsilon_o E^2 A}{2} \quad ...(2)$$

where A is the "effective area" of the plate which is equal to the actual area of the plate A plus half of the area of the gap between A and G substituting E from Eq. (1) Eq. (2), we get

$$F = \frac{\varepsilon_o}{2}\left(\frac{V_a - V_b}{d}\right)^2 A$$

or $$V_a - V_b = d\sqrt{\frac{2F}{\varepsilon_o A}} \text{ Volts} \quad ...(3.1)$$

Procedure for Measuring the Potential Difference

1. At first the plates A, B and C are connected to the earth so that the potential difference between the plates is zero. A small mass m is then *placed* on the plate A, due to weight mg, A gets depressed *below* G with the help of the screw nut N, the plate A is raised to the same place as GG.

2. Now, the small mass m is removed so that the plate A is pulled up by the spring S and is *above the place of G*. The *earth connection* of all the plates is now *broken.*

3. Next the plate *A is connected to a potential* V_0 *i.e.,* the plate is charged to the potential Vo by a machine (Kelvin Replenishes). The plate B is then *connected to one of the points,* the potential difference between which we wish to *measure.* The position of B is adjusted with the help of screw M until the plate A comes in *level with the guard ring.* The reading on the scale is noted, say, R_1 corresponding to the distance between the plates d1. Here, the force of attraction between A and B is *mg.* If the potential of the first point be V_1, then from Eq. (3.1)

$$V_o - V_1 = d_1 \sqrt{\frac{2mg}{\varepsilon_0 A}} \qquad ...(3.2)$$

4. Finally, the plate B is *connected to the second point* whose potential is V_2. The position of B is again in adjusted so that A comes in level with the guard ring. Let the reading on scale be R_2 and the distance between the plates be d_2, then

$$V_o - V_2 = d_2 \sqrt{\frac{2mg}{\varepsilon_0 A}} \qquad ...(3.3)$$

Subtracting (3.3) from (3.2), we get

$$V_2 - V_1 = (d_1 - d_2) \sqrt{\frac{2mg}{\varepsilon_0 A}} \qquad ...(4.4)$$

Since $d_2 - d_1 = R_2 - R_1$, we get

$$V_2 - V_1 = R_1 - R_2 \sqrt{\frac{2mg}{\varepsilon_0 A}} \qquad ...(3.5)$$

Absolute Electrometer

It is called the *"absolute electrometer"* because the potential difference $V_2 - V_1$ in equation (3.5) is known in terms of the difference between the two micrometer readings, the mass m and the area A. That is the *potential difference $V_2 - V_1$ is obtained in terms of "fundamental quantities" such* as mass and length, which can be measured. No calibration is required in this electrometer, hence it is called an absolute electrometer.

It is, however, *less sensitive* than other electrometers. Further, it requires much setting, therefore, it is more cumbersome to use it than a voltmeter.

MEASUREMENT OF DIELECTRIC CONSTANT USING ATTRACTED DISC ELECTROMETER

If the material is given in the form of a slab of thickness, say, t which can be introduced between the plates A and B. The effective air distance between the plates decreases and arrangement is now simply the parallel plate capacitor partially filled with dielectric, its capacity,

$$C = \frac{\varepsilon_0 A}{\left[d - t\left(1 - \frac{1}{k}\right)\right]},$$

where K is the dielectric constant of the slab. Clearly, the effect of introducing of a dielectric slab of thickness *t is effectively decreasing the distance between plates A and B by an amount t* $\left(1 - \frac{1}{K}\right)$ *Hence, if after introducing the slab the separation between the plates is increased by an amount* $h = t\left(1 - \frac{1}{K}\right)$, the capacity of the capacitor reduces to its original value, *i.e.*, $C = \frac{\varepsilon_0 A}{d}$ without the dielectric slab

Hence, $$h = t\left(1 - \frac{1}{K}\right)$$

or $$\frac{h}{t} = 1 - \frac{1}{K}$$

or $$\frac{1}{K} = 1 - \frac{h}{t} = \frac{t-h}{t}$$

or $$K = \frac{t}{t-h} \qquad \text{...(3.6)}$$

Therefore the value of K can be determined in terms of t and h. The value of t and h can be easily determined with the help of attached disc electrometer when the slab is introduced between the plates A and B. The potential is obtained by replacing d in Eq. (3.1) by $d - t\left(1 - \frac{1}{K}\right)$, *i.e.*,

$$V_a - V_b = \left[d - t\left(1 - \frac{1}{K}\right)\right]\sqrt{\frac{2F}{\varepsilon_o A}} \qquad \text{...(3.7)}$$

since $F = \frac{\varepsilon_o E^2}{2}$ and $E = \frac{V_a - V_b}{d - t\left(1 - \frac{1}{K}\right)}$

Now, if the potential difference between the plates is maintained constant and the dielectric slab introduced, the distance between the plates effectively slab introduced, the distance between the plates effectively decreases and electric field E increases so consequently F (force of attracting) increases. Hence, the plate A is pulled down by the plate B by an amount $h = t\left(1 - \frac{1}{K}\right)$ with the help of micrometer screw, the force of micrometer screw, the force of attraction will decrease to its original value and the plate A will return will decrease to its original value and the plate A will return to its original position. Hence h is known from the two readings of micrometer screw.

For measuring the thickness of the slab, the plate B is moved up so that the top and the bottom surfaces of the slab touch the plates A and B respectively. The reading of the position of B is noted. Next, the slab is removed and the plate B is again moved such that B touches A and reading of B is noted again. The difference between two readings is thickness t. Therefore, K is known by Eq. (3.6).

QUADRANT ELECTROMETER

The attracted disc electrometer, though it gives absolute values is not very *sensitive*. Lord Kelvin devised another more sensitive electrometer known as quadrant electrometer. It was further modified by Dolezalek. The Dolezalek quadrant electrometer possesses *high sensitivity and accuracy.*

Construction : The quadrant electrometer consists of a hollow cylindrical box of brass, 5 cms. in diameter and 1 cm in height, divided into four hollow quadrants, The grandaunts are mounted separately on amber insulating stand. The opposite pairs of quadrants AA and BB (Fig. 3.2) are joined together to the terminals T_1 and T_2 respectively can be connected to a source potential Va and Vb. N is a light paddle-shaped van called the meedle which is made of thin paper coated with silver or aluminium. The needle is suspended by a quartz fibre so as to swing freely in a horizontal place without touching the grandaunt. A small mirror M is attached to the suspension fibre to read the deflection by lamp scale arrangement.

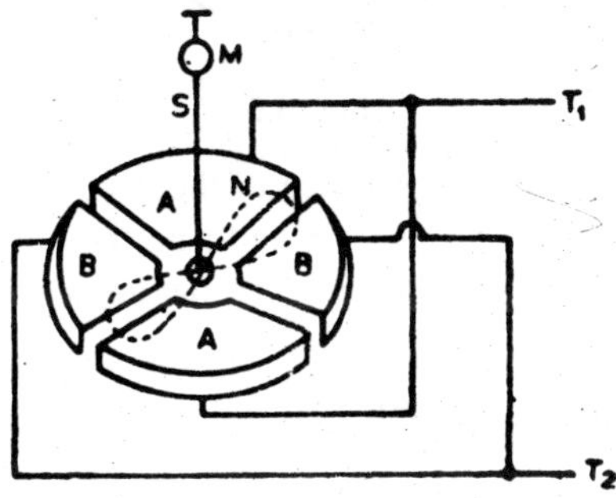

Fig. 3.2

Working : When the pairs of quadrants AA and BB are earthed or connected to the same potential, the needle rests symmetrically between them. But when these are charged to different potentials the needle deflects from the higher to the lower potential. This twists the suspension fibre S and a restoring couple acts which bring the needle to rest. *When the deflecting couple due to the applied potential difference becomes equal to the restoring couple* due to the twist, the resulting deflection of the needle, say, θ is proportional to the difference of potential $V_a - V_b$. This is shown in Fig. 3.3 where dotted line indicates the new position of the needle N and θ is deflection. Thus,

$$V_a - V_b \ \theta \propto$$

The portion of needle, which lies between one pair of the quadrants forms a parallel plate condenser, and that which lies between the other pair forms another condenser. When the needle moves through an angle θ from AA quadrants to BB quadrants, the area of needle shown dotted in Fig. 3.3 is transferred from AA to BB quadrants. Thus when the needle is deflected, its area in quadrants BB increases and that in quadrants AA decreases.

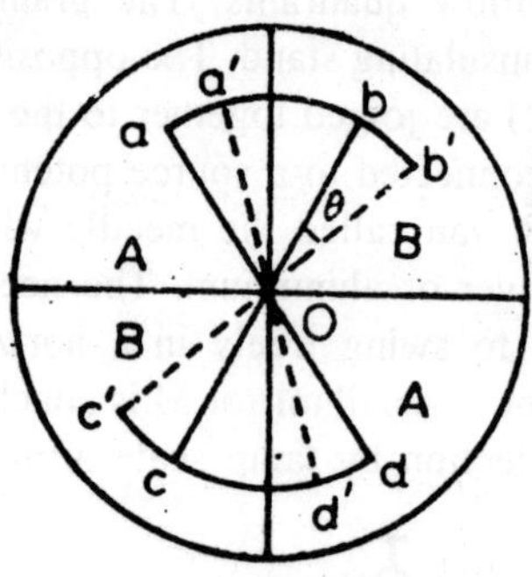

Fig. 3.3

The increase in area of one face of the needle in BB is area Obb' or Occ'. Let area Obb' or Occ' is S and r be the radius of the needle, then an angle 2π represents an area πr^2 the angle θ will represent an area obtained by the condition, *i.e.*,

$$= \frac{\text{area of Obb}'(\text{S})}{\text{area of the circle } \pi r^2}$$

$$= \frac{\text{angle subtended by the Obb' at centre } \theta}{\text{angle subtended by the circle at centre } 2\pi}$$

$$\text{Hence S} = \frac{\theta}{2\pi} \cdot \pi r^2 = \frac{1}{2} r^2 \theta$$

This is same as the corresponding decrease in the area of one face of needle in AA. As the needle has two arms and two faces, therefore, the total area of the needle transferred from AA to BB

$$= 4. \frac{1}{2} r^2 \theta = 2r^2 \theta$$

If d' is the distance between the needle and the quadrant, the *increase in the capacity of the* condenser formed by the needle and quadrants BB due to this increases in area in given by

$$DC = \frac{\varepsilon_o 2r^2\theta}{d} \qquad ...(3.8)$$

since, C = eoA/d,

However, *the capacity of the condenser formed by the needle and quadrants AA will decrease by an "equal amount".* Since the capacity of BN condenser increases while the potential difference $(V_o - V_b)$ between N (needle) and B remains constant, it "*gains charge*" whereas the capacity of AN condenser decreases while the potential difference $(V_o - V_a)$ between N and A remains constant, it "*loses charge*".

Gain of charge by BN condenser = Change in capacitance × p.d.

$$= \Delta C\ (V_o - V_b)$$

"*Gain of electrical energy*" by BN condenser

$$= \text{charge} \times \text{potential difference}$$

$$= \Delta C\ (V_o - V_b).\ (V_o - V_b)$$

$$= \Delta C\ (V_o - V_b)^2$$

Similarly, the "*loss of electrical energy*" by AN condenser

$$= \Delta C\ (V_o - V_a)^2$$

Therefore, the "*net gain of energy*".

$$= \Delta C\ [(V_o - V_b)^2 - (V_o - V_a)^2]$$

$$= \Delta C\ [\{V_o - V_b + V_o - V_a\}\ \{V_o - V_b - V_o + V_a\}]$$

$$= \Delta C\ [(2V_o - (V_a + V_b)]\ (V_a - V_b)$$

$$= 2\Delta C \left[V_o - \frac{V_a + V_b}{2}\right] (V_a - V_b) \qquad ...(3.9)$$

This is to be supplied by the source of electrical energy which is responsible for maintaining the difference of potential $(V_a - V_b)$.

The energy gain given by Eq. (3.9) is used in two parts:

1. In increasing the electrostatics potential energy of the system, and
2. In doing mechanical work in twisting the suspension fibre when the needle is deflected.

The increases in the electrostatics potential energy of the BN condenser

$$= \frac{1}{2}\Delta C (V_o - V_b)^2$$

and the loss of potential energy of the AN condenser is

$$\frac{1}{2}\Delta C (V_o - V_b)^2$$

Therefore, the net gain of potential energy,

$$= \frac{1}{2}\Delta C [V_o - V_b)^2 - (V_o - V_a)^2]$$

$$= \frac{1}{2}\Delta C [2V_o - (V_a + V_b)] (V_a - V_b)$$

$$= \Delta C \left[V_o - \frac{V_a + V_b}{2}\right](V_a - V_b) \qquad ...(3.10)$$

Comparison of equations (3.9) and 3.10 shows that "half" of the energy supplied by the source is used in "*increasing*" the "*electrical potential*" energy of the system. Hence the "*remaining half*" of the energy supplied by the source is used in "*twisting the suspension fibre*" during the deflection of the needle. Hence the energy available for twisting the suspension fibre is

$$= \Delta C \left[V_o - \frac{V_a + V_b}{2}\right](V_a - V_b)$$

Putting the value of ΔC from Eq. (3.8), we get

Energy spent (or work done) in twisting the fibre

$$= \frac{2\varepsilon_o r^2\theta}{d}\left[Vo - \frac{V_a + V_b}{2}\right](V_a - V_b)$$

But the work done = twisting couple × angle of twist

Therefore, twisting or "*deflecting couple*"

$$= \frac{2\varepsilon_o r^2\theta}{d}\left[Vo - \frac{V_a + V_b}{2}\right](V_a - V_b)$$

If τ is the restoring couple per unit twist, then the couple for the deflection θ is

$$\tau\theta = \frac{2\varepsilon_0 r^2}{d}\left[V_0 - \frac{V_a + V_b}{2}\right](V_a - V_b)$$

or $$\theta = \frac{2\varepsilon_0 r^2}{d\tau}\left[V_0 - \frac{V_a + V_b}{2}\right](V_a - V_b) \quad ...(3.11)$$

Thus, the deflection θ is proportional to $\frac{2\varepsilon_0 r^2}{d\tau}$ which is constant, to the potential deference $(V_a - V_b)$ between AA and BB and to the average difference of potential between the needle and quadrant $\left(\text{i.e.} V_0 - \frac{V_a - V_b}{2}\right)$.

The quadrant electrometer can be used in two ways:

(1) Electrostatic use, and

(2) Idiostatic use.

(1) Heterostatic Use

In this case the needle is charged to a very high potential as compared to the potentials Va and Vb, which are given to the quadrants AA and BB respectively as shown in Fig. 3.4 (a). Thus, when

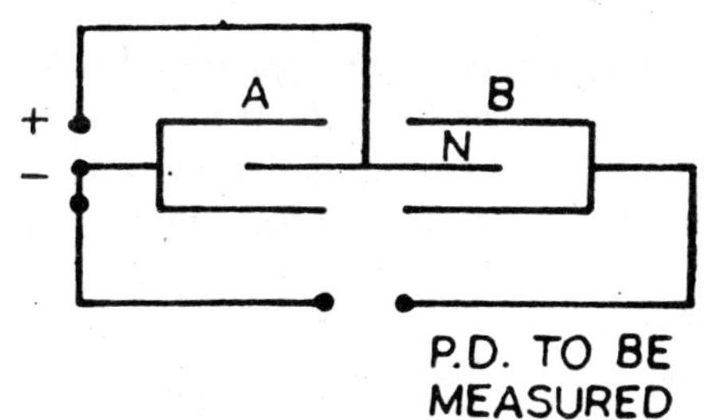

Fig. 3.4(a)

$$V_0 > V_a \text{ or } V_b$$

Then $\frac{V_a + V_b}{2}$ may be neglected in Eq. (3.11) and hence,

$$\theta = \frac{2\varepsilon_0 r^2}{d\tau} V_0 (V_a - V_b)$$

or $$\theta = K\,(V_a - V_b) \qquad \text{...(3.12)}$$

where, $K \dfrac{2\varepsilon_o r^2}{d\tau} V_o$

Fig. 3.4(b)

Therefore, *the deflection in the heterostatic use is proportional to the potential difference.* If a graph is plotted between deflection and potential difference a straight line is obtained as shown in Fig. 3.4(b).

Limitation of This Method

Since $V_o > V_a$ or V_b, *this method is applicable for the measurement of the "small potential difference" only.* If the potentials applied to two pairs of quadrants are reversed the deflection will also be reversed, *hence in this use electrometer can measure only the "direct voltage"*

(2) Idiostatic Use

In the case, the needle and one pair of quadrants, say AA are connected together as shown in Fig. 3.5(a), hence,

$V_o = V_a$ and Eq. (3.11) reduces to

$$\theta = \frac{\varepsilon_o r^2}{d\tau} (V_a - V_b)^2$$

$$= K' (V_a - V_b)^2 \qquad \text{...(3.13)}$$

where, $K' = \dfrac{\varepsilon_o r^2}{d\tau}$

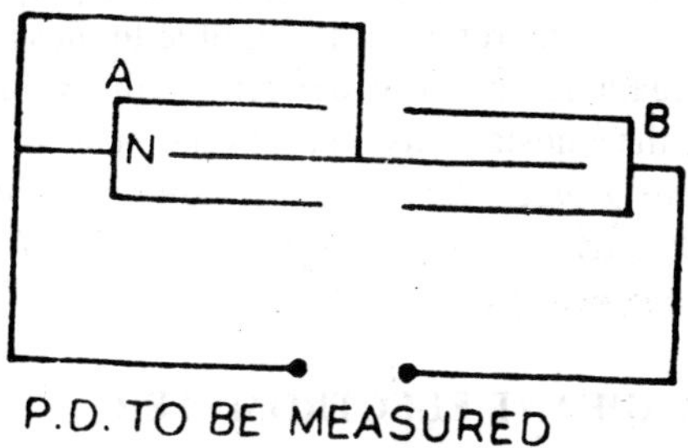

Fig. 3.5(a)

Thus, *the deflection in the idiostatic use is proportional to the "square of the potential difference"*. Hence, *this method is used for determining " alternating potential difference"* and *"large DC potential"* when the meedle and quadrants AA are connected to the higher terminal. The variation of deflection with p.d. is shown in Fig. 3.5(b)

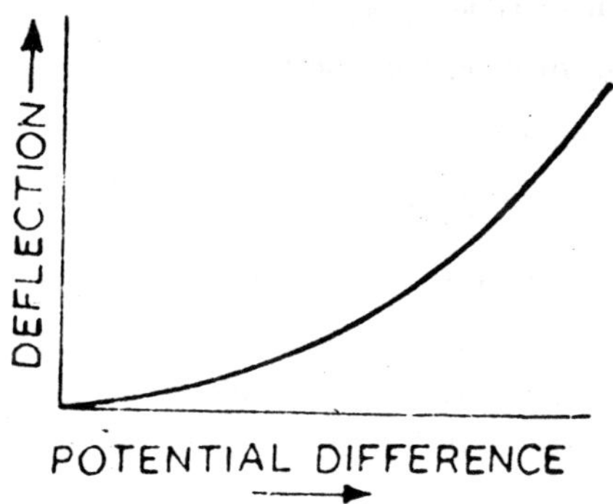

Fig. 3.5(b)

Relative Merits of Heterostatic and Idiostatic Uses

We have, $\dfrac{K}{K'} = \dfrac{2\varepsilon_0 r^2}{d\tau} V_0 \Big/ \dfrac{\varepsilon_0 r^2}{d\tau}$

$= 2V_0$ ...(3.14)

Hence K > K'. Thus, *the electrometer when used electrostatically is more accurate than when used idiostatically because in the heterostatic use, we have comparatively larger deflection for the same difference of potential.* But in idiostatic use the deflection is proportional to the square of the potential, therefore, it is possible to measure alternating potential also. Again for a given deflection, the value $V_a - V_b$ is much greater on the idiostatic use than in the heterostatic use, since K' < K, hence the *range of the instrument is much more in the idiostatic use, although the accuracy is considerably less* in this case than to heterostatic use.

USES OF A QUADRANT ELECTROMETER

A quadrant electrometer can be used for the following purposes:

1. To measure small direct potential difference (or voltage). For details, see heterostatic use discussed above.
2. To measure alternating potential difference. For detail see idiostatic use discussed above.
3. Measurement of capacity and comparison of two capacities
4. Measurement of capacity current.
5. Measurement of dielectric constant.
6. Comparison of potentials.

Important cases are discussed here separately as follows:

Measurement of Capacity

If one pair of the quadrants is earthed and the other is connected to a potential V, then the deflection of needle given by equations (3.12) *i.e.,*

$$\theta = K (V_a - V_b)$$

here, $V_b = 0$ and $V_a = V$

hence, $\theta = KV$ ...(3.15)

This equation is used for the measurement of the capacity of a quadrant electrometer and for the comparison of two capacities.

(a) Capacity of a Quadrant Electrometer

The circuit arrangement is shown in Fig. 3.6. Let C be the capacity of the electrometer, one pair of quadrants of the electrometer is charged to a constant V_1 with a standard cell S by depressing the key K' and the resulting deflection θ_1 is noted. Then according to equation (3.15), we have,

$$\theta_1 = KV_1$$

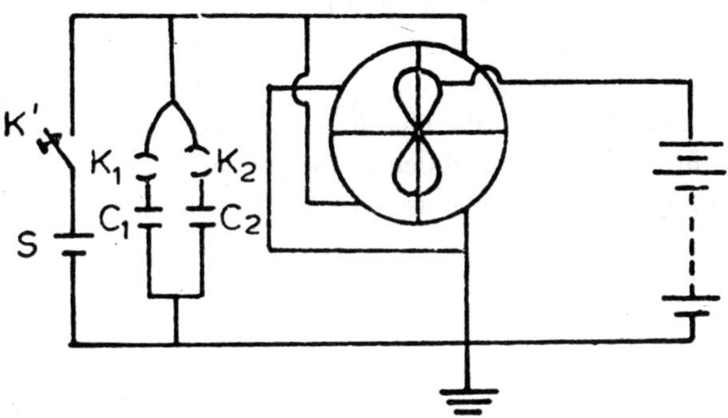

Fig. 3.6

Now K' is released and the standard condenser C_1 is put in circuit by closing the key K_1. The condenser C_1 shares the charge of the electrometer and the common potential becomes V_2 which gives the deflection θ_2 given by,

$$\theta_2 = K\Delta_2$$

then the total charge on the electrometer Q is given by,

$$Q = CV_1 = (C + C_1)\, V_2$$

or $$\frac{C + C_1}{C} = \frac{V_1}{V_2} = \frac{\theta_1}{\theta_2}, \left(\text{Since} \frac{KV_1}{KV_2} = \frac{\theta_1}{\theta_2}\right)$$

or $$\frac{C_1}{C} = \frac{\theta_1 - \theta_2}{\theta_2}$$

or $$C = C_1 \frac{\theta_1}{\theta_1 - \theta_2} \qquad ...(3.16)$$

Thus C can be calculated. A guard-ring capacitor is preferred as a standard condenser C_1 because its capacity can be calculated from its geometrical dimensions, as,

$$C_1 = \frac{\varepsilon_0 A}{d}$$

(b) Comparison of Two Capacities

In this case, first the condenser C1 put in the circuit by closing the key K_1 and then the key K' is depressed so that both the capacity C_1 and the electrometer are charged to a common potential V. Let θ be the corresponding deflection in electrometer. Further, the key K_1 is opened and the condenser C_2 is put in the circuit by closing the key K_2, C_2 shares some of the charge acquired by C_1 and electrometer as shown in Fig. 3.6. Consequently the potential falls to V' and the corresponding deflection becomes q', then the charge,

$$Q = (C_1 + C)\,V = (C_1 + C_2 + C)\,V'$$

or
$$\frac{C_1 + C_2 + C}{C_1 + C} = \frac{V}{V'}$$

But V Kθ and V' = Kθ'

hence,
$$\frac{C_1 + C_2 + C}{C_1 + C} = \frac{\theta}{\theta'}$$

or
$$\frac{C_2}{C_1 + C} = \frac{\theta - \theta'}{\theta} \qquad \text{...(3.17)}$$

Since C, the capacity of the electrometer is known form equation (3.16), the ratio C_2/C_1 becomes known, however, if C is very small in comparison to C_1 and C_2, we can neglect C in equation (3.17),

Then,
$$\frac{C_2}{C_1} = \frac{\theta - \theta'}{\theta'} \qquad \text{...(3.18)}$$

Determination of Dielectric Constant

As dielectric constant K is defined as,

$$K = \frac{\text{Capacity of condenser with the subs tan ce as dielectric}}{\text{Capacity of the condenser with air as dielectric}}$$

Hence the methods of determination of capacity of capacity and dielectric constant are same. Any method used for comparison of two capacities can be used for determining dielectric constant.

Measurement of Feeble (Ionisation) Currents

A quardant electrometer can be used to measure "very feeble ionisation current." The ionisation current is produced by exposing a chamber containing gas, by some ionising agent, e.g. X-rays, ultraviolet or by radiations from radioactive materials.

As ionisation chamber, consisting of a metallic box containing two metal plates A and B, insulated and separated from each other. The plate A is charged to a high positive potential and to one pair of quadrants. The other plate B is connected to one pair of quadrants of the electrometer, which is earthed. The needle of the electrometer is given a high positive potential as shown in Fig. 3.7.

When the gas in the chamber is ionised, the needle of the electrometer is deflected because due to ionisation the negative ions are produced and they move toward the electrode A. *The movement of the ions constitutes ionisation current. The rate of change of deflection is proportional to the magnitude of the ionisation current.*

Let C be the capacity of the electrometer together with the electrode A to which it is connected, and if V be the potential at any instant t due to the charge Q, the ionisation current,

I = rate of flow of charge

$$= \frac{dQ}{dt} = \frac{CdV}{dt}$$

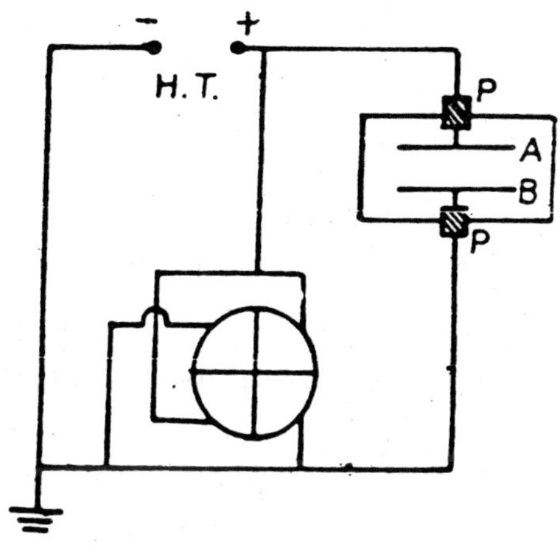

Fig. 3.7

And from equation (3.15), θ = KV, where K is constant.

$$\text{Hence, } I = C\frac{V_1 - V_2}{t} \qquad \text{...(3.19)}$$

$$\text{or} \quad I = \frac{1}{K}C\frac{d\theta}{dt} = K'\frac{Cd\theta}{dt} \qquad \text{...(3.20)}$$

where K' = 1/K another constant.

$\frac{d\theta}{dt}$ is the rate of change of deflection of the needle, C is determined by experiment described earlier and constant K' = 1/K, is determined from the calibration graph. Hence I can be calculated. Currents of the order of 10^{-15} amp. can be measured with a quadrant electrometer.

Measurement of Dielectric Constant

(i) Hopkinson's Null Method for Solids

We can determine the dielectric constant by this method if the solid is available in the form of slab. The circuit arrangement is shown in Fig. 3.8, C_2 is a guard ring condensers which the lower plate is movable. C_1 is a variable cylindrical condenser consisting of co-axial cylindrical conductors, the inner cylinder can slide inside the outer one, hence the capacity of this condenser can be increased or decreased by increasing or decreasing the area of overlap. The battery B is earthed at its mid-point so that its terminal points are at equal and opposite potentials with respect to the earth-point. The needle of the electrometer is charged by battery E to a high positive potential. When the two keys K_1 and K_2 are depressed, the condensers C_1 and C_2 are charged such that the upper plate of C_1 is at a positive potential and that of C_2 to an equal negative potential. Hence, if two condensers are of equal capacity, they will acquire equal and opposite charges. Now, if the keys K_1 and K_2 are raised such that they are in contact with C_1 and C_2 and the connection with battery is broken, the charges will neutralize one another and there will be no deflection in the electrometer. This is obtained by adjusting the capacity of the variable capacitor C_1.

Let the distance between the plates A and B of the capacitor C_2 be d. The given dielectric slab of thickness t is placed on the plate B. This increases the capacity of C_2. The previous adjustment of C_1 is kept undisturbed and the palet B is moved downward until there is again no deflection in the electrometer when K_1 and K_2 are raised such that C_1 and C_2 are connected to electrometer. Let x be the distance by which the plate B is moved.

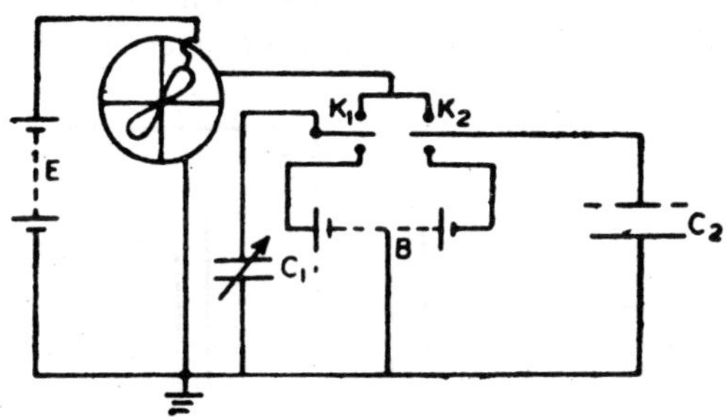

Fig. 3.8

Capacity of C_2 without dielectric $= \frac{\varepsilon_0 A}{d}$ where A is the effective area of the guard ring capacitor. Capacity of C_2 after introducing the dielectric slab, when the electrometer shows no deflection,

$$= \frac{\varepsilon_0 A}{\left[d + x - t\left(1 - \frac{1}{K}\right)\right]}$$

Therefore, $$\frac{\varepsilon_0 A}{d} = \frac{\varepsilon_0 A}{\left[d + x - t\left(1 - \frac{1}{K}\right)\right]}$$

or $$d = d + x - t\left(1 - \frac{1}{K}\right)$$

or $$K = \frac{t}{t - x} \qquad ...(3.21)$$

Hence K can be calculated.

(ii) Dielectric Constant of a Liquid by Hopkinson's Method

Procedure is the same as that for solid except that the guarding condenser C_2 in Fig. 3.8 is replaced by a double-walled cylindrical condenser in which an insulated cylinder C could hang as shown in Fig. 3.8. This condenser is connected in place of C_2 in circuit of Fig. 3.8 and C_1 is suitably adjusted, so that there is no deflection of the needle of the electrometer. Then the capacity of condenser C is equal to the capacity of the condenser C_1.

If a and be are the inner and outer radii of the cylindrical condenser C_1 and ll is the length of overlap of inner and outer cylinders.

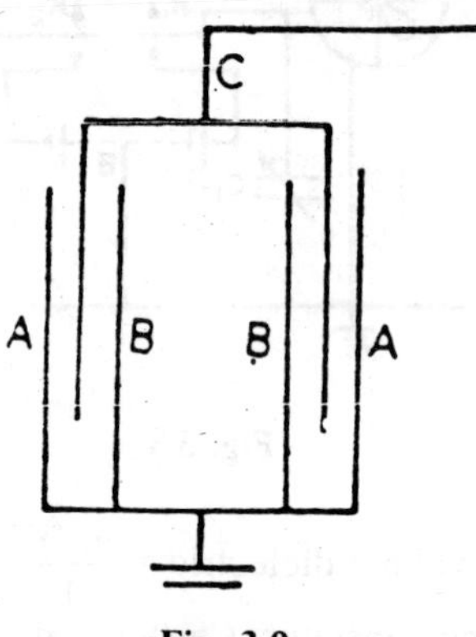

Fig. 3.9

Then the capacity of the condenser C, is equal to the capacity of C_1 without dielectric.

$$C_{\text{without dielectric}} = \frac{2\pi\varepsilon_0 l_1}{\log_e \frac{b}{a}}$$

Now the given liquid is filled in C between A and B and the position of C_1 is again adjusted for zero deflection. If l_2 is the length of overlap then capacity of the condenser C with dielectric,

$$C_{\text{with dielectric}} = \frac{2\pi\varepsilon_0 l_2}{\log_e \frac{b}{a}}$$

the ratio, $$\frac{C_{\text{with dielectric}}}{C_{\text{without dielectric}}} = \frac{l_2}{l_1} = K \qquad ...(3.22)$$

Hence, K, *i.e.*, dielectric constant, or the relative permittivity of the liquid, is known.

(iii) Boltzmann's Method of Determining the Dielectric Constant of Gases

The circuit used by Boltzmann is shown in Fig. 3.10. Two plates P and Q forming a parallel plate condenser are enclosed in a metal chamber C which is earthed, the chamber C one be exhausted or filled with gas. The connections from the plates P and Q are insulates from the chamber C. The plate P is connected to the positive of H.T. battery of n cells each of voltage V though key K_1. The plate Q is connected to one pair of the quadrants AA and the other quadrants are earthed. The key K_2 helps to connect plate Q and quadrants AA to the earth.

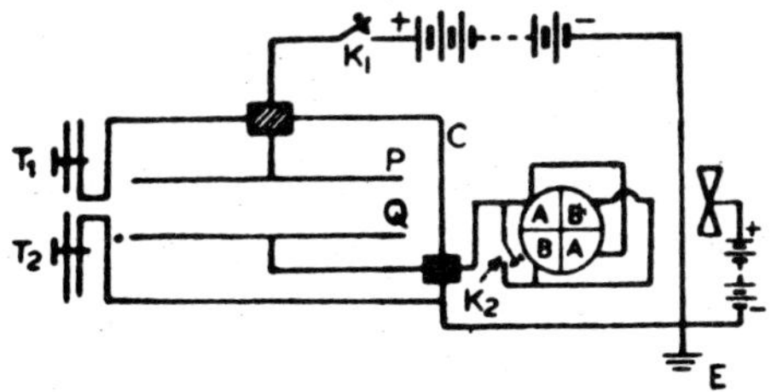

Fig. 3.10.

The chamber C is first evacuated then the key K_1 is closed which charges the plate P to a potential nV. If the key K_2 is simultaneously closed as K_1, there will be no deflection in the electrometer because both the Paris of quadrants are earthed. Next, keys K_1 and K_2 are opened and the chamber is filled with given gas. The capacity of the condenser formed by P and Q will increase K times and consequently, the potential of P falls to nV/K.

The key K_1 is again closed and the plate P is charged to a potential nV. It receives additional charge from the source due to its increased capacity. This additional charge induces an equal and opposite charge on Q. Thus, Q is charged by the same amount as that of P. As Q is connected to electrometer, hence electrometer needle is deflected. Let this deflection be θ, then according to eq. (3.15),

$$\theta \propto V$$

Hence the change in potential $= nV - \dfrac{nV}{K}$

$$= nV\left(1 - \frac{1}{K}\right)$$

Therefore, $\theta \propto nV\left(1 - \dfrac{1}{K}\right)$

or $$\theta = K'\, nV\left(1 - \frac{1}{K}\right) \qquad ...(3.23)$$

where K' is a constant.

Now if one more cell is added to H.T. battery then the potential of P will increase to nV + V = (n + 1) V and the potential of Q will be also increased by an equal amount. This increase of potential changes the deflection of the needle say θ', then,

$$\theta' \propto V$$

or $$\theta' K' V \qquad ...(3.24)$$

Hence, $$\frac{\theta}{\theta'} = n\left(1 - \frac{1}{K}\right)$$

or $$1 - \frac{1}{K} = \frac{\theta}{n\theta'}$$

or $$\frac{1}{K} = 1 - \frac{\theta}{n\theta} = \frac{n\theta' - \theta}{n\theta'}$$

or $$K = \frac{n\theta'}{n\theta' - \theta}$$

Thus knowing n, θ and θ', K can be calculated.

Comparison of Potentails

Potentials can be compared either by connecting the electometer electrostatically or idiostatically and using eq. (3.12) and (3.13) respectively.

ELECTROSTATIC VOLTMETER

The electrostatic voltmeter was first designed by Lord Kelvin and is a modified form of quadrant electrometer. It consists of only one pair of opposite quadrants AA placed vertically. A light aluminium needle N is pivoted at the centre so that it can rotate about a horizontal axis in a vertical place symmetrically between the quadrants. The needle N is insulated from the quadrants. The upper part of the needle carries a pointer which moves over the scale S of the instrument and the lower

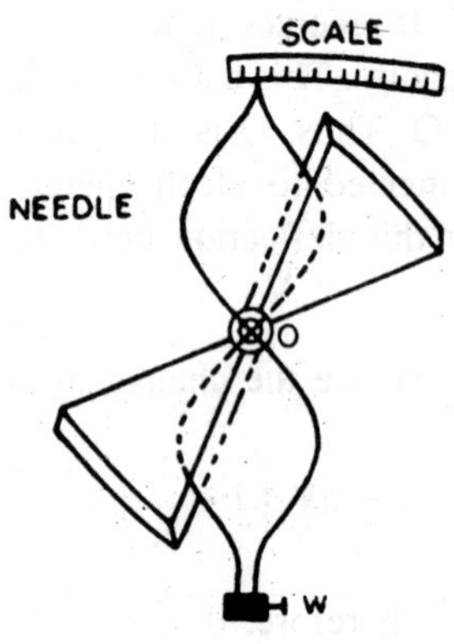

Fig. 3.11

part carries a controlling weight which can be adjusted for changing the range of the instrument as shown in Fig. 3.11.

The potential difference to be measured is applied between the needle and the quadrants. The needle N rotates due to this potential difference and the deflection of the needle is proportional to the square of the applied potential difference. The scale is directly calibrated to read volts by actual comparison with the standard cell or attracted disc electrometer. Thus, the device is handy and is capable of measuring potential of several thousand volts.

ELECTROSTATICS MACHINES

The devices used to produce electrical charges on a large scale are called electrostatics machines or generators. In fact, *these machines do not generate the charges but they only separate the charges.* In general, there are two types of machines namely,

(1) Frictional machines,

(2) Induction machines.

Now a days the frictional machines are not used.

The commonly used induction machines are:

(1) Electrophorus, divested by Volta in 1775.

(2) Wimshurst machines,

(3) Van de Graaff generator

Here, we shall discuss only the Van de Graaff generator, for other *induction machines see some school books.*

VAN DE GRAAFF GENERATOR

This was designed by Van de Graaff in 1931. It can produce high voltage of the order of 50 mega volt. The principle of the generator is as follows.

It uses the *discharge action of points and charge collecting action of hollow spherical conductors.* It consists of large hollow sphere S mounted on a large pillar P of highly insulating material. B is an endless belt of some insulating material, *i.e.,*., rubber or silk as shown in Fig. 3.12. Belt runs on motor driven pulleys P_1 and P_2 and enters the sphere through the slit S_1 and leaves the sphere through S_2. A conductor whose one end is a point A, is charged to a high potential (about 10,000 volts) is placed near the belt. This charge leaks out to the belt. This charge leaks out to the belt which becomes positively charged. The belt moves up and passes to another close point H near P_2 connected to the metallic spheres S.

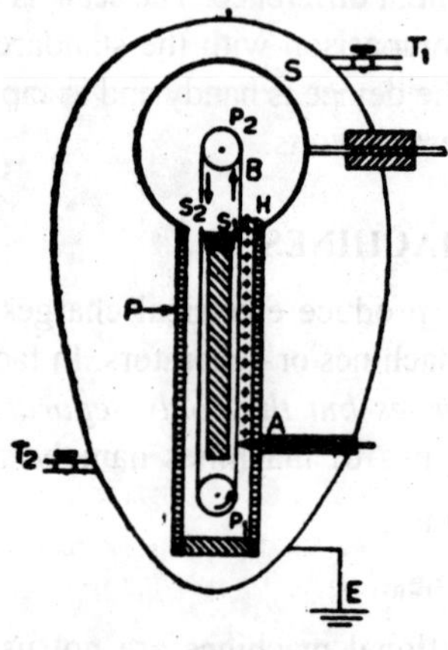

Fig. 3.12

This induces a negative charge on the point H and positive charge on the sphere. The negative charge on the point H is soon neutralised by the positive charge on the belt. The sphere thus gets positively charged. As the belt is revolving continuously, the charge on the sphere is accumulating contiguously. *Consequently the potential of the sphere rises considerably depending on its size and the smoothness of its surface. The bigger the sphere and the smoother its surface, the greater is the potential it can retain without leakage.*

To prevent the discharge taking place from S, the generator is completely enclosed in an earthed steel tank which is filled with air under pressure.

SOLVED EXAMPLES

Example 1:

Calculate the force of attraction between the plates of an attracted disc electrometer when a potential of 3,000 volts is applied between them, they are separated by distance of 0.5 cm and area of each plate is 10 sq. cm.

Solution:

Potential difference = 3,000 volts

Electric field

$$\text{Force} = -\frac{\varepsilon_o E^2 A}{2}$$

$$= \frac{8.85\times10^{-12}}{2}\frac{\text{Coul}^2}{\text{Nt}\times\text{m}^2}\times10\times10^{-4}\,\text{m}^2\left(\frac{3000}{05\times10^{-2}}\frac{\text{Volts}}{\text{m}}\right)^2$$

$$= \frac{8.85\times10^{-12}\times3\times3\times10^{6}\times10\times10^{-4}\,\text{Coul}^2}{2\times25\times10^{-6}}\frac{\text{Coul}^2}{\text{Nt}}\left(\frac{\text{Volt}}{\text{m}}\right)^2$$

$= 157.3 \times 10^{-5}$ Newtons, since Volt = Nt. m/Coul.

Example 2:

Find the mass, required to give a force equal to the force of attraction between the plates of attracted disc electrometer when they are 4 mm apart, having area 10 sq. cm and at a potential difference of 1000 volts.

Solution:

$V_2 - V_1 = 1{,}000$ volts

and $d_1 - d_2 = 4 \times 10^{-3}$ m

and $A = 10 \times 10^{-4}\ \text{m}^2$

Using Eq. (3.4), *i.e.*,

$$V_2 - V_1 = d_2 - d_1\sqrt{\frac{2mg}{\varepsilon_o A}}$$

on putting the values,

$$1000 = 4 \times 10^{-3}\sqrt{\frac{2\times9.8\times m}{8.85\times10^{-12}\times10^{-3}}}$$

or $m = 2.8 \times 10^{-5}$ kg

Example 3:

Two large metal plates are fixed horizontally at a distance 0.5 cm from each other. What potential in volts should be applied *between the plates, if a droplet of oil of mass 1.5×10^{-11} gms and carrying a charge 4.9×10^{-10} e.s.u. is to be held at rest between the plates.*

Solution:

Let the potential to be applied be V volts, then electric field,

$$E = \frac{V}{0.5 \times 10^{-2}} \text{ Volts/m}$$

Upward force on the droplet

$$qE = \frac{V}{0.5 \times 10^{-2}} \times \frac{4.9 \times 10^{-10}}{3 \times 10^{9}} \text{ Newtons}$$

and this is balanced by weight, mg,

$$= 1.5 \times 10^{-14} \times 9.8 \text{ Newtons}$$

$$\text{Hence, } \frac{V}{0.5 \times 10^{-2}} \cdot \frac{4.9 \times 10^{-10}}{3 \times 10^{9}} = 1.5 \times 10^{-14} \times 9.8$$

or $\quad V = 4500$ volts

Example 4:

An electrometer is charged by connecting it momentarily to a battery and a deflection of 100 division is observed. Then a condenser of 50 μF capacitance is connected in parallel to the electrometer and deflection falls to 75 divisions. Calculate the capacity of the electrometer.

Solution:

Let C be the capacity of the electrometer and C_1 is the capacity of the condenser. In the first case,

$$Q = CV$$

and in second case,

$$Q = (C + C_1) V_1$$

or

$$\frac{V}{V_1} = \frac{C + C_1}{C}$$

But from equation (3.15), *i.e.*, $\theta = KV$

$$\frac{V}{V_1} = \frac{\theta}{\theta_1} = \frac{100}{75}$$

therefore,

$$\frac{C + c_1}{C} = \frac{4}{3}$$

or $$\frac{C_1}{C} = \frac{1}{3}$$

or $$C = 3C_1 = 3 \times 50\ \mu F$$

$$= 150\ \mu F$$

Example 5:

Two leyden jars are exactly similar in size and shape but one has glass as dielectric and the other ebonite. The glass jar is charged by when the charge is shared between the two jars the potential falls to 0.6 of its original value. If dielectric constant of ebonite is 2, find that of glass.

Solution:

Using Eq. (3.18),

$$\frac{C_1}{C_2} = \frac{\theta'}{\theta - \theta'} \qquad \text{(but } \theta = KV\text{)}$$

hence, $$\frac{C_1}{C_2} = \frac{V'}{V - V'}$$

Let V be the original potential, the,

$$V' = \frac{6}{10} V$$

Therefore, $$\frac{C_1}{C_2} = \frac{6V/10}{V - \frac{6V}{10}} = \frac{3}{2}$$

Since the capacitors are exactly similar their capacitance without dielectric will be same. Hence,

$$C_1 = K_1\ C_o$$

where K_1 is dielectric constant of glass

and $$C_2 = K_2 C_0$$

where K_2 is dielectric constant of ebonite.

Then,

$$\frac{C_1}{C_2} = \frac{K_1}{K_2} = \frac{3}{2}$$

or $$K_1 = \frac{3}{2} K_2$$

$$K_{glass} = 3$$

(as $K_2 = 2$)

Example 6:

When the air inside an electroscope of 20 cm capacitance is ionised by a beam of X-rays, it is found that the potential of the gold leaf changes from 154 to 100 volts in one minute. Find the current (1 farad = 9×10^{11} cm).

Solution:

$$C = 20 \text{ cm} = \frac{20}{9 \times 10^{11}} \text{ farad}$$

and $$I = \frac{CdV}{dt} = \frac{20}{9 \times 10^{11}} \times \frac{54}{60} = 2 \times 10^{-11} \text{ amp.}$$

EXERCISES

1. Calculate the force of attraction between lower and upper discs of an attracted disc electrometer when a potential difference of 1000 volts is applied between them. Given that they are 0.5 cm apart and area 10 sq. cm.

2. Two circular metal plates each of radius 2 cm are separated by a distance of 2mm in air. Calculate the potential difference in volts between them when the force of attraction between them is equal to one gm weight.

3. Two brass plates are arranged horizontally 2 cm apart and the lower plate is earthed. The plates are charged to a potential difference of 400 volts. A drop of oil carrying a charge 4.8×10^{-10} esy is at rest between the plates. Calculate the mass of drop.

4. A small current flows through a resistance of 10^{10} ohms, the ends of which are connected to opposite quadrants of an

electrometer. The deflection is 120 scale divisions. When a denial cell (emf 1.04 volts) is connected across these quadrants, the deflection is 54 scale divisions. Calculate the current.

5. Give the principle and theory of the attracted disc electrometer.
6. (a) Why is it called absolute electrometer?
   (b) What advantage has it over other forms of electrometer?
   (c) How would you find dielectric constant of a substance with attracted disc electrometer?
   (d) What are the advantages and disadvantages of an electrometer over a voltmeter?
7. Give the theory of quadrant electrometer. Explain clearly the heterostatic and idiostatic arrangements and discuss the relative advantages.
8. Describe the quadrant electrometer and explain how it may be used to measure
   (1) Alternating potential
   (2) Direct potential
   (3) Ionisation current
9. Give the theory of quadrant electrometer. How will you use it to compare two capacities?
10. What is Van de Graaff generator? Discuss its principle and working.

Multiple Choice Questions

1. Electometer is an accurate device because
   (a) High current is required for its working
   (b) Low current is required for its working
   (c) The current completely avoided
   (d) None of the above.
2. Absolute measurement of potential difference is done by
   (a) Gold leaf electroscope
   (b) Quadrant electrometer
   (c) Kelvin's attracted disc electrometer
   (d) Voltmeter.

3. Kelvin's absolute electrometer is less sensitive than other electrometers. This statement is

   (a) False

   (b) True.

4. The dielectric constant can be measured by electrometer. This is

   (a) True

   (b) False.

5. Quadrant electrometer is more accurate when used.

   (a) Iodiostatically

   (b) Heterostatically.

6. For measuring the alternating potential difference the quadrant electrometer must be used

   (a) Iodiostatically

   (b) Heterostatically.

7. For measuring small potential difference the quadrant electrometer is used

   (a) Iodiostatically

   (b) Heterostatically.

8. For measuring large potential difference the quadrant electrometer is used

   (a) Idiostatically

   (b) Heterostatically.

9. Measurement of feeble ionisation current is possible by

   (a) Ammeter

   (b) Quadrant electrometer.

10. A quadrant electrometer can be used to measure

    (a) Capacitance

    (b) Inductance

    (c) Potential difference

    (d) None of the above.

4

Capacitors and Dielectrics

CAPACITOR

A capacitor (or condenser) is a device for storing charge. In actual practice, a capacitor is an electrical device consisting of two conductors separated by an insulating or dielectric medium (including air) and carrying equal and opposite charges. The conductors are called pates and may be of any shape. The earliest capacitor was invented almost accidentally by VAN MUSSCHENBROEK of Leden in about 1746, and became known as a Leden jar.

PARALLEL PLATE CAPACITOR

As an example let us take two similar flat conducting plates arranged parallel to one another, separated by distance d, as shown in Fig. 4.1. Let A be the area of each plate and suppose that there is a charge Q on one plate and –Q on the other. Away from the edge, the field $\vec{E}$ is very nearly uniform in the region between the plates and given by,

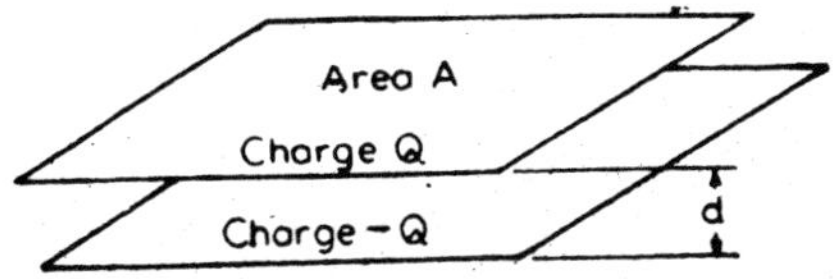

Fig. 4.1

$$E = \frac{\sigma}{\epsilon_0} \quad \text{or} \quad \sigma = \epsilon_0 E$$

where σ is the density of the surface charge on the inner surface of the plate. If V is the potential difference between the plates then, $E = V/d$, hence above equation reduces,

$$\sigma = \epsilon_0 \frac{V}{d}$$

If we neglect the variation of $\vec{E}$ near the edge of the plates and therefore of σ, the total charge on one plate represented by Q is given by,

$$Q = \epsilon_0 \frac{VA}{d}$$

or
$$\frac{Q}{V} = \frac{\epsilon_0 A}{d} \qquad ...(4.1)$$

The ratio $\frac{Q}{V} = \frac{\epsilon_0 A}{d}$, *though it is only approximation in parallel plates, but it is clear that the value of R.H.S. will depend only on the size and geometrical arrangement of the plates.* That is, *for a fixed pair of conductors, the ratio of charge to potential difference will be a constant. We call this constant the capacitance of the capacitor and denoted by C, hence,*

$$\frac{Q}{V} = \text{Constant} = C$$

or
$$Q = CV \qquad ...(4.2)$$

Thus, the capacity of the parallel plate capacitor, with edge fields neglected, is given by,

$$C = \frac{\epsilon_0 A}{d} \qquad ...(4.3)$$

Units

From Equation (4.2) unit of capacitance is coul/volt and is called Farad related by,

$$1 \text{Farad} = \frac{1 \text{ coulomb}}{1 \text{ volt}} \qquad ...(4.4)$$

However, farad is very large unit, we use smaller unit as micro farad or micro-micro farad.

$1\ \mu F = 10^{-6}$ F and $1\ \mu\mu F = 10^{-12}$ F.

Types of Capacitors According to Shape

The value of C given in Eqn. (4.3) depends on the size and geometry of the plates, hence we can design the capacitors of different geometry, the more commonly used capacitors are,

1. a parallel plate capacitor
2. a cylindrical capacitor
3. a spherical capacitor

A CYLINDRICAL CAPACITOR

Such a capacitor consists of two coaxial cylinder of radius a and b and length l. Calculate capacitance of device, sectional diagram is given in Fig. 4.2. Construct a Gaussian surface shown by dotted line having radius r, by using Gauss's law we get,

$$\epsilon_0 \oint \vec{E}.d\vec{S} = Q$$

where Q is the charge on the surface.

$$\epsilon_0 \; E.2\pi rl = Q$$

$$E \frac{Q}{2\pi \; \epsilon_0 \; rl} \text{ at P}$$

the potential difference between the plates is given by,

$$V = -\int_b^a \vec{E}.d\vec{r}$$

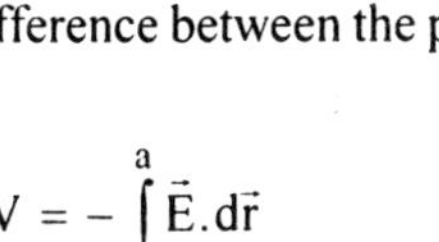

Fig. 4.2

As the angle between $\vec{E}$ and $d\vec{r}$ is 0, since $\vec{E}$ is radial *i.e.*, along $\vec{r}$, so $\vec{E}.d\vec{r}$ = Edr, Hence,

$$V = -\int_b^a E\,dr = -\int_b^a \frac{Q}{2\pi \; \epsilon_0 \; rl} dr$$

$$= \frac{Q}{2\pi \; \epsilon_0 \; l} \log_e \frac{b}{a}$$

Therefore, the capacitance,

$$C = \frac{Q}{V} = \frac{2\pi \; \epsilon_0 \; l}{\log_e (b/a)} \qquad ...(4.5)$$

Like (4.3) *equation, the capacitance C given in equation* (4.5) *also depends on geometrical factors.*

SPHERICAL CAPACITORS

Let us calculate the capacitance of two concentric spheres shown in Fig. 4.3. The potential difference between the two spheres V, given by,

$$V = -\int_b^a \vec{E}.d\vec{r}$$

$\vec{E}$ at a distance r is given by,

$$\vec{E} = \frac{Q}{4\pi \in_0} \frac{\hat{r}}{r^2}$$

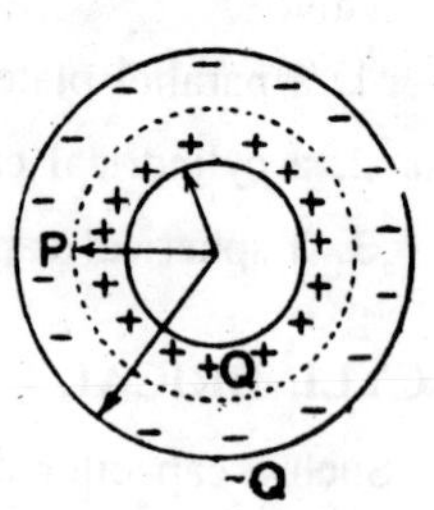

Fig. 4.3

$\vec{E}$ is radial *i.e.*, along r, hence angle between $\vec{E}$ and $\vec{r}$ is zero. Therefore, $\vec{E}.d\vec{r}$ = Edr. Hence,

$$V = -\int_b^a \frac{Q}{4\pi \in_0} \frac{dr}{r^2} = -\frac{Q}{4\pi \in_0}\int_b^a \frac{dr}{r^2}$$

$$= \frac{Q}{4\pi \in_0}\left(\frac{1}{a} - \frac{1}{b}\right) \text{ and } C = 4\pi \in_0 \frac{ab}{b-a} \qquad ...(4.6)$$

Here again the capacitance C depends only on their radii.

GUARD RING CAPACITOR

To overcome the edge effect in parallel plate capacitor the two rings one on right and other on left side are attached as shown in Fig. 4.4. These rings are called guard ring to ensure uniformity in field and such capacitor is called guard ring capacitor.

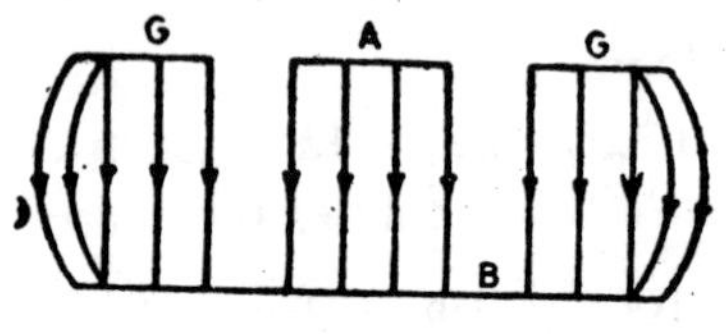

Fig. 4.4

ENERGY STORED IN A CAPACITOR

The energy of a charged capacitor is the amount of work done in charging it. Suppose V is the potential difference between the plates of the capacitor of capacitance C. The charge Q is given by,

$$Q = CV$$

There is charge Q on one plate and — Q on the other. Let us increase the charge from Q to Q + dQ by transferring a positive charge dQ from the negative to the positive plate, working against the potential difference V. The work that has to be done is given by,

$$dW = VdQ = \frac{Q}{C}dQ$$

Therefore, to charge the capacitor starting from the uncharged state to the final charge say, Q

$$W = \int_{Q=o}^{Q=Q} \frac{Q}{C}dQ = \frac{1}{2}\frac{Q^2}{C} = \frac{1}{2}CV^2 = \frac{1}{2}QV \qquad ...(4.7)$$

This work done is stored up in the form of potential energy and supplies the energy necessary to drive the current when the capacitor is discharged.

For a parallel plate capacitor the C is given by,

$$C = \frac{\epsilon_o A}{d} \text{ and } V = Ed$$

Hence the energy stored is given by,

$$U = \frac{1}{2}\frac{\epsilon_o A}{d} E^2.d^2 = \frac{1}{2}\epsilon_o E^2.A.d \text{ joules} \qquad ...(4.8)$$

Energy stored per unit volume

$$= \frac{1}{2}\epsilon_o E^2 \text{ joules/metre}^2 \qquad ...(4.9)$$

(A.d is the volume between the plates). This expression for energy is same as the general formula for the energy in an electric field given in Eqn. (4.41).

FORCE OF ATTRACTION BETWEEN CAPACITORS PLATES

Consider a simple parallel plate capacitor such as illustrated in Fig. 4.5. It appears reasonable, since one plate is positively charged and the second plate is negatively charged, that there should be a force of attraction between the plates tending to pull them together.

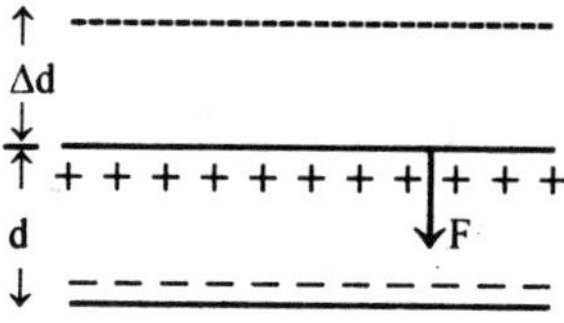

Fig. 4.5

An expression for the force can be readily obtained by the application of the *principle of virtual work.* Suppose E is intensity of electric field between the plates. The total energy in the field is given by,

$$U = \frac{1}{2} \epsilon_0 E^2 Ad \text{ joules}$$

Let us now calculate the change in the total energy when the capacitor plate to be separated by an additional amount Δd. *The work that would be required to effect the additional separation is equal to the change in energy in the field,* and this, must be equal to the product of the force and distance. The energy per unit volume is given by,

$$U = \frac{1}{2} \epsilon_0 E^2 Ad \text{ joules/metre}^3$$

The change in volume due to Δd change in separation = A Δd. Hence, the change in total energy due to change in volume

$$= \frac{1}{2} \epsilon_0 E^2 A \Delta d \qquad ...(4.10)$$

The mechanical work done from the outside against the force of attraction between the plates in moving them further apart through a distance Δd will be,

$$W = \vec{F}.\Delta\vec{d} \qquad ...(4.11)$$

or $$W = -F \Delta d \qquad ...(4.12)$$

This work done = change in energy

$$\frac{1}{2} \epsilon_0 E^2 A \Delta d = \vec{F}.\Delta\vec{d}$$

or $$\frac{1}{2} \epsilon_0 (\vec{E}.\vec{E}) A \Delta d = \vec{F}.\Delta\vec{d} = F \Delta d$$

The *negative sign indicates that the force acting over the area A is normal to the plates in the direction of* E, *i.e.,* F *and* Δd *are* 180° apart. The magnitude is given.

$$F = \frac{1}{2} \epsilon_0 E^2 A \qquad ...(4.13)$$

But the force per unit area is pressure, and hence the pressure is given by,

$$P \text{ (Pressure)} = \frac{F}{A} = \frac{1}{2} \epsilon_0 \vec{E}\vec{E}$$

or $$P = \frac{1}{2} \epsilon_0 E^2 \text{ newton/ metre}^2 \qquad ...(4.14)$$

This expression for pressure is the same as the general formula for pressure on a conductor due to electric field $\vec{E}$ discussed in eq. (4.37).

THE DIELECTRIC CONSTANT

Now *we will discuss, insulator or dielectric materials which do not conduct electricity*. However, using a simple electroscope and two parallel plate capacitors Faraday discovered that this was not so. He constructed two identical capacitors in one of which he placed a dielectric. When both capacitors were charged to the same potential difference as shown in Fig. 4.6a.

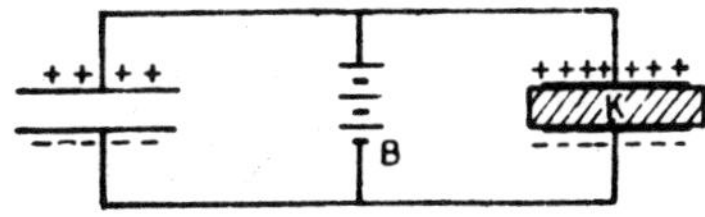

Fig. 4.6(a)

Faraday found by experiment that the charge on the capacitor with dielectric is greater than that without. *Since* q *is larger, for the same V, it follows from the relation* C = q/V *that the capacitance of a capacitor increases if dielectric is placed between the plates.* If dielectric material fills completely, the space between the plates *the ratio of the capacitance* C *with the dielectric to that without* C_0 *is called dielectric constant,* K, *which is the property of dielectric material i.e.,*

$$C/C_0 = K \qquad ...(4.15)$$

K is also *called relative permittivity* of the medium.

Faraday obtained the same relation Eq. (4.15) by placing same charge to the capacitors as in Fig. 4.6b and measured the potential

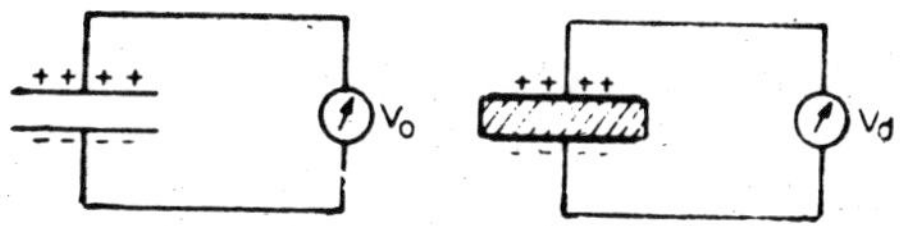

Fig. 4.6(b)

difference. If V_d is the potential difference with dielectric and V_o without dielectric capacitor then $V_d = \frac{V_o}{K}$. So the potential difference with dielectric decreases by a factor 1/K. Hence from relation C = q/V as q is same, here again the same conclusion.

CLASSIFICATION OF DIELECTRIC ON THE BASIS OF MOLECULAR STRUCTURE (POLAR AND NON POLAR MOLECULES)

The molecules or atoms are composed of electrons and nuclei in a state of vibration about certain fixed points in space. Though the molecule or atom as a whole is neutral *i.e.,* the magnitude of charge on electrons is equal to the positive on nuclei due to protons. But the spacial arrangement of charges in a molecule may be different in various substances. All positive and negative charges of a molecule can be replaced by an equivalent single positive and single negative charge located at the centres of gravity of separate positive and separate negative charges respectively. These centres of gravity can be found with the aid of the same rules which are used to determine the position of the centres of gravity of systems of point masses in statics. These centres of gravity may either coincide or not coincide in space.

At any given instant, if the centre of gravity of the positive and negative electrical charges "do not coincide", then the molecule is "polar" and the matter composed of such molecules is also called polar. When the two centres of gravity are separated, the molecules may momentarily possess a dipole moment equal to the product of charge and distance of separation. While if the two centres of gravity "coincide." then the molecules is called a "non-polar molecule" having negligible dipole moment and the matter composed of such molecules is also called non-polar. Matter comprising different molecules both non-polar and polar should be regarded as polar matter.

Examples of Non-Polar Molecules

The rigid molecules which have symmetrical structure are non-polar and have always as negligible small dipole moment. They are

H_2, N_2, Cl_2, CH_4, CCl_4, C_6H_6, C_2H_2

and C_2H_4.

Example of Polar Molecules

Most molecules are, however, polar, there average dipole moment is finite. *If symmetry is disturbed,* the molecule becomes polar. If a hydrogen atom of methane molecule is replaced by chlorine atom, it losses symmetry and becomes polar.

A DIELECTRIC IN AN ELECTROSTATIC FIELD (POLARISATION OF DIELECTRIC)

We shall discuss this in the next section in detail, here is an *elementary atomic view. A dielectric whether polar or nonpolar will acquire dipole moment by induction when placed in an electric field.* Then these elementary dipoles are oriented along the field direction, dielectric becomes polarised without affecting the neutrality of dielectric. In other words, *by induction the one side get positive charge while other the same negative charge.* So the electric field setup by them (E') opposes the external electric field say E_0. Fig 4.7 shows the induction process. *One important point to be noted is that induced effect is present only when the electric field is present.* The resultant field $\vec{E} = \vec{E}_0 + \vec{E}'$. The field E' due to induced charge points opposite to E_0, hence the field inside dielectric decreases same as potential. So we can write, using the relation,

$$E = Vd$$

$$\frac{V_0}{Vd} = \frac{E_0}{E} = K \qquad ...(4.16)$$

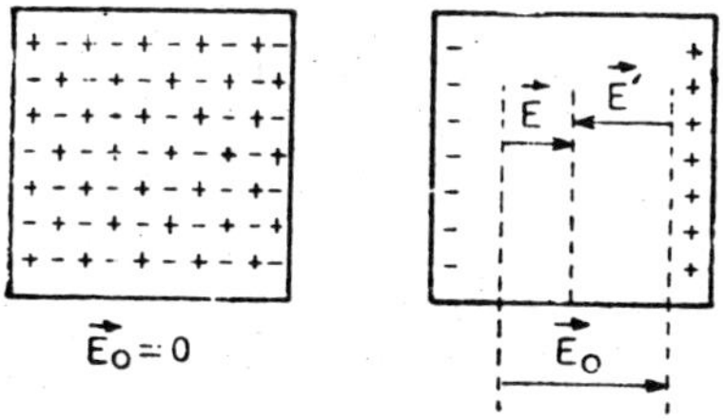

Fig. 4.7

CAPACITY OF A PARALLEL PLATE CAPACITOR PARTIALLY FILLED WITH DIELECTRIC

Fig. 4.8 shows a capacitor partially filled with a dielectric material of thickness t. If Q is the charge on each plate of area A. Then value of electric field E_o due to charge Q in the free-space between plates and dielectric material given by Gauss's law

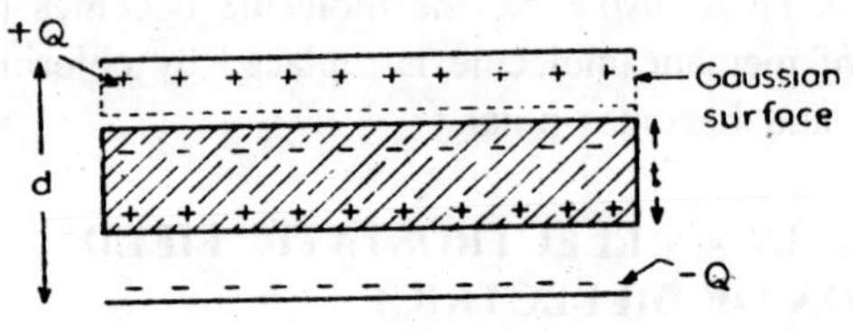

Fig. 4.8

$$\epsilon_0 \oint \vec{E}.d\vec{S} = \epsilon_0 E_0 A = Q$$

or $$E_0 = \frac{Q}{\epsilon_0 A} \quad ...(4.17)$$

From Eq. (4.14) the field E inside the dielectric is given by,

$$E \frac{E_0}{K} = \frac{Q}{K \epsilon_0 A} \quad ...(4.18)$$

E_o is the value of electric field in the free space (*i.e.,* without dielectric) which is dt, and E is the value of electric field inside the dielectric of thickness t. The potential difference between the two points is given by,

$$(V = V_1 - V_2 = -\int_2^1 \vec{E}.d\vec{l})$$

Hence the total potential difference between the plates A and B,

$$V_a - V_b = -\underset{\substack{\text{for}\\\text{free}\\\text{space}\\d-t}}{\int \vec{E}_0.d\vec{l}} - \underset{\substack{\text{for}\\\text{thick}\\\text{ness}\\t}}{\int \vec{E}.d\vec{l}}$$

Since the total field is $(\vec{E}_o + \vec{E})$

Therefore,

$$V_a - V_b = Et + E_o (d - t)$$

$$(\vec{E}.d\vec{l} = -Edl, \text{similarly, } \vec{E}_0.d\vec{l} = -E_0 dl)$$

Substituting E_o and E from equation (4.17) and (4.18), we get,

$$V_a - V_b = \frac{Q}{A \in_0}\left(d - t + \frac{t}{k}\right)$$

Therefore, from equation (4.1) the capacitance C is given by,

$$C = \frac{Q}{V_a - V_b} = \frac{\in_0 A}{\left[d - t\left(1 - \frac{t}{k}\right)\right]} \quad (.4.19)$$

DIELECTRICS AND GAUSS'S LAW

Now apply Gauss's law to a parallel plate capacitor filled with a dielectric of dielectric constant K. Fig. 4.9a shows an usual parallel plate capacitor without dielectrics. Let charge be Q on the plates with proper sign and A be the area of each plate. E_0 is the electric field without dielectric.

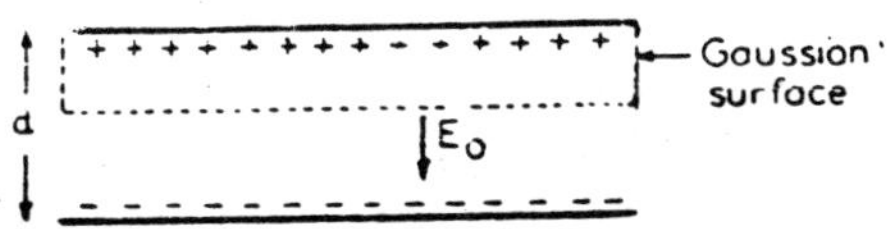

Fig. 4.9(a)

Therefore, the Gauss's law gives,

$$\in_0 \oint \vec{E}.d\vec{S} = \in_0 E_0 A = Q$$

or

$$E_0 = \frac{Q}{\in_0 A}$$

Now place the dielectric as shown in Fig. 4.9b then the net charge enclosed by Gaussian surface is Q – Q'. Therefore with dielectric the Gauss's law is,

Fig. 4.9(b)

$$\in_0 \oint \vec{E}.d\vec{S} = Q - Q' \qquad ...(4.20)$$

or $$\in_0 EA = Q - Q'$$

or $$E = \frac{Q}{\in_0 A_0} - \frac{Q'}{\in_0 A} \qquad ...(4.21)$$

in which $-$ Q' is the induced charge due to polarisation. Using the Equation (4.16), we have

$$E = \frac{E_0}{k}$$

where E is field inside the dielectric substituting E_0, we have,

$$E = \frac{Q}{k \in_0 A} \qquad ...(4.22)$$

Substituting this value of E in Eq. (4.21), we have,

$$\frac{Q}{k \in_0 A} = \frac{Q}{\in_0 A} - \frac{Q'}{\in_0 A} \qquad ...(4.23)$$

or $$Q' = Q\left(1 - \frac{1}{k}\right) \qquad ...(4.24)$$

If there is no dielectric between the plates then K = 1, and Q' = Q, *i.e., there will be no induced charge.* The value of K is always greater than one (K > 1), hence Q' (induced charge) is always less than Q (free charge).

From Equation (4.24) the value of Q – Q' is given by,

$$Q - Q' = \frac{Q}{k}$$

Now we can write the Gauss's law with dielectric as given in equation (4.20) after substituting Q – Q' in the following form, *i.e.,*

$$\in_0 \oint \vec{E}.d\vec{S} = Q - Q'$$

or $$= \frac{Q}{k}$$

or $$\in_0 \oint K\vec{E}.d\vec{S} = Q \qquad ...(4.25)$$

Though we have derived equation (4.25) for a parallel plate capacitor, but it is applicable in all cases when the dielectrics are present.

CAPACITY OF A PARALLEL PLATE CAPACITOR WHEN COMPLETELY FILLED WITH DIELECTRIC

Fig. 4.9b represents such capacitor. In such circumstances, the value of intensity of electric field is E between the plates is given by,

$$E = \frac{Q}{k \in_0 A}$$

and thickness of the slab is same as the separation between the plates *i.e.,* d. Hence,

$$V = Ed$$

$$V = \frac{Qd}{k \in_0 A}$$

therefore, $$C = \frac{Q}{V} = \frac{k \in_0 A}{d} \text{ Farad} \qquad ...(4.26)$$

ELECTRIC POLARISATION VECTOR P

The polarisation P is defined as the induced surface charge per unit area i.e., surface density of bound charges in dielectric,

$$P = \frac{Q'}{A} = \sigma \text{ pol} \qquad ...(4.27)$$

The induced charge Q' appears when the dielectric is polarised that is the reason we call Q'/A as electric polarisation P.

Now multiply the numerator and denominator of Eq. (4.27), by d the thickness of the dielectric slab we get,

$$P = \frac{Q'd}{Ad}$$

The product Q'd is the induced electric dipole moment because by polarisation the dielectric slab develops the charge Q' on one surface and —Q' on the opposite separated by thickness of the slab, hence a dipole of dipole moment Q'd. The product A.d is the volume (v) of the slab therefore, we can write,

$$P = \frac{\text{Induced dipole moment}}{\text{Volume}} = \frac{Q'd}{v} \qquad ...(4.28)$$

The polarisation P is defined as the induced dipole moment per unit volume. Since the electric dipole moment is a vector the electric polarisation is also a vector and its magnitude is P.

$$\vec{P} = \frac{Q'\vec{d}}{v} \qquad ...(4.29)$$

Thus the magnitude P is the polarisation charge per unit area on the surface, *i.e.*, the surface density of charges (σ_{pol}) appearing at right angles to the direction of the applied field. Hence,

$$\sigma_{pol} = \frac{Q'}{A} = p \qquad ...(4.30)$$

The equation (4.30) can be generalised by considering the case when dielectric surface is not perpendicular to P.

Let the normal to the surface OO' make an angle θ with the direction of P as shown in Fig. 4.10. The component of P normal to the surface is equal to $\vec{p}.\hat{n}$, where $\hat{n}$ is a unit vector along the normal to the surface, we have,

$$\sigma\ pol = \vec{p}.\hat{n} \qquad ...(4.31)$$

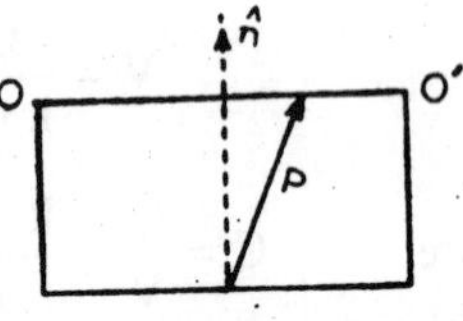

Fig. 4.10

$\vec{p}$ at a Point

The value of P in equation (4.28) is an average for the volume Ad. To define the meaning of p at a point, it is convenient to assume that a dielectric in an electric field has *continuous distribution of infinitesimal dipoles,* that is a continuous polarisation. Let us consider a continuously polarised dielectric, *the value of* $\vec{p}$ *at a point can be defined as the net dipole moment* $\vec{p}$ *of a small volume* Δ v *divided by the volume,* with the limit taken a Δv shrinks to zero around the point, thus,

$$\vec{P} = \lim_{\Delta v \to 0} \frac{\vec{P}}{\Delta v} \qquad ...(4.32)$$

UNIFORM AND NON-UNIFORM POLARISATION

When isotropic and homogeneous materials (*i.e.*, the materials having constant value of K hence the constant value of dipole moment or polarisation vector $\vec{p}$) are placed in an electric field developed opposite charges of equal magnitude of the two sides of the surface because of polarisation. *Hence the net charge or average charge density is zero.* Such polarisation is called "uniform polarisation". The charge density remains the positive and, negative charges are displaced relative to each other. Their displacement does not produce any net charge inside the volume.

On the other hand, if the material is *non-isotropic, and in homogeneous,* K and $\vec{p}$ are not constant, say $\vec{p}$ is large at one place and smaller at another. This means that more charge be moved into some region than away from it. We then get a *volume density of* charge.

The total charge displaced out of any volume v by the polarisation is the integral of the outward drawn normal component of P over the surface that bounds the volume, see Fig. 4.11.

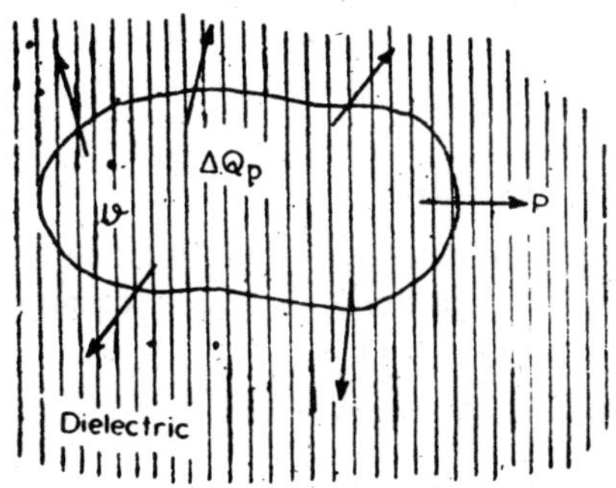

Fig. 4.11

An equal charge of the opposite sign is left behind when charge crosses the surface. Let the net charge enclosed by volume be ΔQ_p. We can write,

$$\Delta Q_p = -\int_S \vec{\sigma}_p . d\vec{S}$$

Now using relation, $\sigma_{pol} = \vec{p}.\hat{n}$, see equation (4.31)
Hence,

$$\Delta Q_p = -\int_S \vec{P}.\hat{n}dS$$

If ρ_p is the volume density of polarised charge in the dielectric the,

$$\Delta Q_p = \int_V \rho_p \, dv$$

equating these two values, we get,

$$\int_V \rho_p \, dv = -\int_S \vec{P}.\hat{n}dS$$

using Gauss's divergence theorem, i.e,

$$\int_S \vec{P}.\hat{n}\,dS = \int_V \text{div}.\vec{P}\,dv = \int_V \vec{\nabla}.\vec{P}\,dv$$

Hence above relation reduces,

$$\int_V \rho_p\,dv = -\int_V \vec{\nabla}.\vec{p}.dv$$

or $$\vec{\nabla}.\vec{p} = -\rho_p \qquad ...(4.32a)$$

In uniform polarisation process which occurs in isotropic materials the value of $\vec{p}$ *is constant throughout the material, hence* $\nabla\vec{p}$. *is zero.* Now let us consider the *non-uniform polarisation.* $\vec{p}$ is not constant, it varies from point to point inside the dielectric material, than $\vec{p}$ is function of position and $\nabla.\vec{p}$ *is given by equation* (4.32 a).

THREE ELECTRIC VECTORS

Let us consider the polarisation of a dielectric slab placed between the plates of parallel plate capacitor and consider equation (4.23), *i.e.,*

$$\frac{Q}{K\epsilon_0 A} = \frac{Q}{\epsilon_0 A} - \frac{Q'}{\epsilon_0 A}$$

or $$\frac{Q}{A} = \epsilon_0\left(\frac{Q}{K\epsilon_0 A}\right) + \frac{Q'}{A} \qquad ...(4.33)$$

According to equation (4.21),

$$\frac{Q}{K\epsilon_0 A} = E$$

and from equation (4.27),

$$\frac{Q'}{A} = P$$

After substitution, equation (4.33) reduces,

$$\frac{Q}{A} = \epsilon_0 E + P \qquad ...(4.34)$$

Put the ratio, $$\frac{Q}{A} = D \qquad ...(4.35)$$

We call it the electric displacement an equivalent term for flux density. As E and P on the right hand side of equation (4.34) are vectors, D is also a vector, hence we can write,

$$\vec{D} = \epsilon_0 \vec{E} + \vec{P}, \text{ everywhere} \quad ...(4.36)$$

or
$$\vec{D} = \left(\epsilon_0 + \frac{\vec{P}}{\vec{E}}\right)\vec{E} \quad ...(4.37)$$

The *displacement vector* is an important addition which is of *great use in Maxwell electro-magnetic equation,* to explain the displacement current. However, from equation (4.35), it is clear that $\vec{D}$ is associated only with free charges.

Fig. 4.12 shows the representation of $\vec{D}$, $\epsilon_0\vec{E}$ and $\vec{P}$ for a parallel plate capacitor with dielectric.

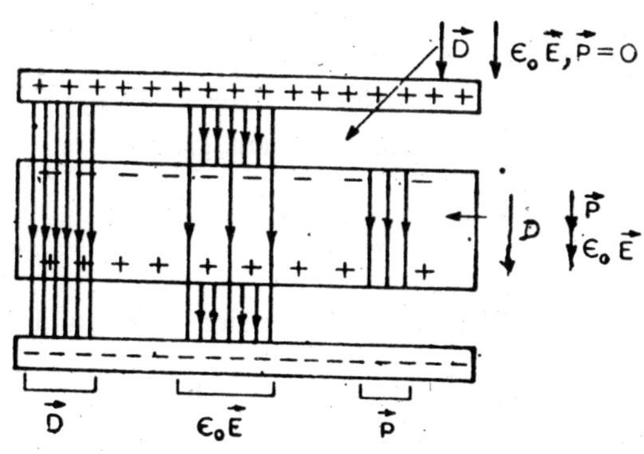

Fig. 4.12

Be careful to note the following points :

(1) $\vec{D}$ *is connected with the free charge only.* The lines of $\vec{D}$ begin and end on free charges. $\vec{D}$ is constant throughout the capacitor.

(2) $\vec{P}$ *is connected with polarisation charge only.* The line of $\vec{P}$ begin and end on polarisation charges. Clearly $\vec{P}$ is zero except inside the dielectric.

(3) $\vec{E}$ *is connected with all charges.*

(4) *In isotropic media* $\vec{P}$ and $\vec{E}$ are in the same direction. However for, *nonisotropic media* $\vec{P}$ and $\vec{E}$ *are, in general, not in the same direction.*

(5) $\vec{E}$ is reduced inside the dielectric, where there are fewer lines.

(6) Unlike the electric field

(which is the force acting on unit charge) or the polarisation $\vec{P}$ (the dipole moment per unit volume), the electric displacement $\vec{D}$ has no clear physical meaning. The only reason for introducing it is that it enables one to calculate fields in the presence of dielectrics without first having to know the distribution of polarisation charges.

Units of $\vec{D}$ and $\vec{P}$

From equations (4.27) and (4.36) the units for $\vec{P}$ and $\vec{D}$ are cous/meter^2 whereas $\vec{E}$ is measured in Newtons/coul.

$\vec{D}$ and $\vec{P}$ in Terms of $\vec{E}$

The vectors $\vec{D}$ and $\vec{P}$ are separately connected to $\vec{E}$. Let us start from the definition of $\vec{D}$, the magnitude of $\vec{D}$ is given by,

$$D = \frac{Q}{A} = K \in_0 \left(\frac{Q}{K \in_0 A} \right)$$

But $\frac{Q}{K \in_0 A} = E$, the expression for $\vec{D}$ can be written as,

$$\vec{D} = K \in_o \vec{E} = \in \vec{E}, \text{ (as } \in = K \in_o) \qquad ...(4.38)$$

Hence relation $\vec{D} = \in_o \vec{E} + \vec{P}$ can be written as,

$$\in \vec{E} = \in_o \vec{E} + \vec{P}$$

or
$$\in = \left(\in_0 + \frac{\vec{P}}{\vec{E}} \right) \qquad ...(4.39)$$

Similarly, we can express $\vec{P}$ in terms of $\vec{E}$ the magnitude of P is given by equation (4.27) *i.e.*,

$$P = \frac{Q'}{A}$$

The induced charge Q' due to polarisation is given in equation (4.24) *i.e.*,

$$Q' = Q \left(1 - \frac{1}{K} \right)$$

Substituting this value in the above equation, we have,

$$P = \frac{Q}{A} = \left(1 - \frac{1}{K} \right)$$

Since, $\frac{Q}{A} = D$, using eq. (4.38), we get,

$$\vec{P} = K \epsilon_0 \vec{E}\left(1 - \frac{1}{K}\right)$$

or $$\vec{P} = \epsilon_0 (K - 1)\vec{E} \qquad ...(4.40)$$

In vacuum K = 1, therefore the polarisation vector $\vec{P}$ is zero.

RESTATEMENT OF GAUSS'S LAW OR (DIVERGENCE OF $\vec{D}$)

The definition of $\vec{D}$ given by equation (4.38) when substituted in equation (4.25), then the simple modified form of the Gauss's law in the presence of dielectric is written as,

$$\oint_S \vec{D}.dS = Q \qquad ...(4.41)$$

or in words, flux of $\vec{D}$ out of a closed surface S = Total free charge enclosed within S.

That is the surface integral of the normal component of electric displacement vector $\vec{D}$ *over any closed surface equals the charge enclosed, where Q is the free charge and the polarisation charge is not included. Equation (4.41) can be* written in divergence form as follows. Since according to divergence theorem,

$$\int_S \vec{D}.dS = \int_V \text{div}\,\vec{D}\,dv$$

and $$Q = \int_V \rho_f\, dv$$

hence, $$\vec{\nabla} . \vec{D} = \rho f \qquad ...(4.42)$$

where ρf is the volume density of free charge Q. For physical significance see the Gauss's law, equation (4.42) is still really Gauss's law, now modified in such a way that the effects of polarisation charge are automatically included.

VERIFICATION OF RELATION $\vec{D} = \epsilon_0 \vec{E} + \vec{P}$

The variation of electric displacement $\vec{D}$ electric field intensity $\vec{E}$, polarisation $\vec{P}$ along the axis between the plates of an air capacitor is shown in Fig. 4.13a. The variation after a dielectric (paraffin) slab has been introduced is illustrated by Fig. 4.13b. For a given applied potential *i.e.,* battery is disconnected when the paraffin is introduced. *In air capacitor equation (4.36) reduces,*

$$\vec{D} = \epsilon_0 \vec{E} \qquad ...(1)$$

because there is no polarisation, hence,

$$\vec{P} = 0, \text{ and } \vec{E} = \frac{\vec{D}}{\epsilon_0} \qquad ...(2)$$

In paraffin capacitor the relation is given in equation (4.36),

$$\vec{D} = \epsilon_0 \vec{E} + \vec{P}$$

The relation between $\vec{D}$ and $\vec{E}$ for this case is given in eq. (4.38).

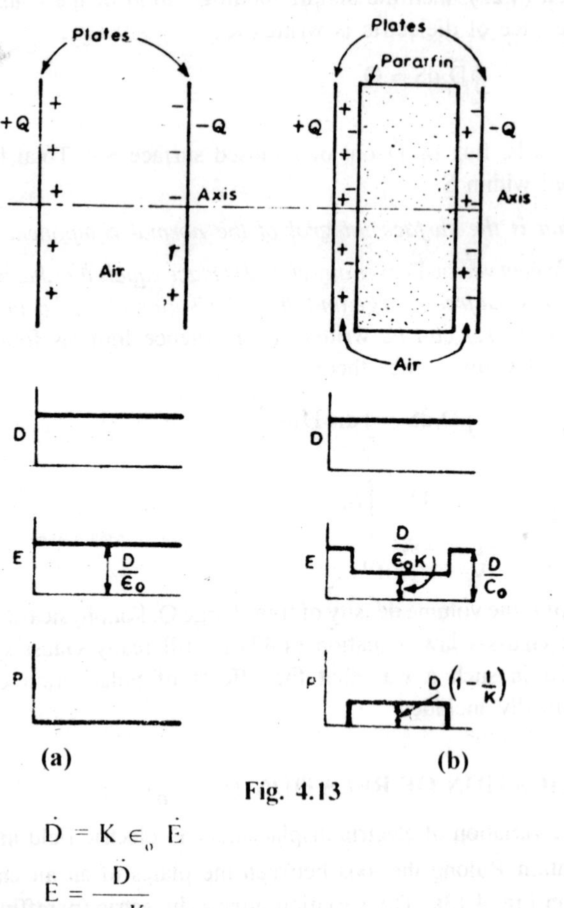

Fig. 4.13

$$\vec{D} = K \epsilon_0 \vec{E}$$

$$\vec{E} = \frac{\vec{D}}{\epsilon_0 K} \qquad ...(3)$$

and the polarisation $\vec{P}$ by equation (4.40),

$$\vec{P} = \epsilon_0 (K - 1)\ \vec{E}$$

or $$\vec{P} = \epsilon_0 K\left(1-\frac{1}{K}\right)\vec{E}$$

$$= \vec{D}\left(1-\frac{1}{K}\right) \quad ...(4)$$

K for paraffin is 2.1, hence the capacitance with paraffin increases by 2.1 times. That is with paraffin dielectric the capacitor plates can store 2.1 times charge for a given potential. *Since the battery was disconnected before the paraffin was introduced, the charge on the plates is same as with air.* Therefore, $\vec{D}$ is the same as before. However, now electric intensity,

$$\vec{E} = \frac{\vec{D}}{\epsilon_0 K} = 0.475\ \frac{\vec{D}}{\epsilon_0}, \text{with air it is } \frac{\vec{D}}{\epsilon_0} \quad ...(5)$$

the polarisation $\vec{P} = \vec{D}\left(1-\frac{1}{K}\right) = 0.525\ \vec{D}$, with air $\vec{P} = 0$ ...(6)

However, the polarisation $\vec{P}$ has such value that

$\vec{D}$ with dielectric $= \vec{P} + \epsilon_0\ \vec{E} = 0.525\ \vec{D} + 0.475\ \vec{D} = \vec{D}$ ...(7)

$\vec{D}$ with air $= \epsilon_0\ \vec{E} = \epsilon_0 \frac{\vec{D}}{\epsilon_0} = \vec{D}$

Hence, equation (4.36) is verified.

DIELECTRIC SUSCEPTIBILITY AND PERMITTIVITY

In most cases (in the so-called linear dielectrics) the magnitude of polarisation is directly proportional to the intensity or the electric field at a given point of a dielectric *i.e.*, we can have,

$$P \propto E$$

or $$\vec{P} = \epsilon_0\ \chi_{er}\ \vec{E} \quad ...(4.43)$$

Substituting the dimensions in equation (4.43), as,

$$P = \frac{\text{couls}}{\text{meter}^2},\ E = \frac{\text{Newtons}}{\text{Coul}},$$

$$\epsilon_0 = \frac{\text{couls}^2}{\text{Newton / meter}^2}\text{m0}$$

We see that χ_{er} is a *dimensionless parameter of dielectric called relative dielectric susceptibility.* The total proportionality factor $\chi_{er} \in_o = \chi e$ has the dimensions Farad/meter is known as absolute *susceptibility or dielectric susceptibility i.e.,*

$$\chi_{er} = \frac{\chi_e}{\in_0} \quad ...(4.44)$$

$$\left[\chi_{er} \in_0 = \frac{P}{E} = \text{coul}^2/\text{Newton. meter}^2\right.$$

$$\text{since volt} = \frac{\text{Newton meter}}{\text{Coul}} \text{ and Farad} = \text{coul/volt}$$

$$\text{Farad} = \frac{\text{coul}^2}{\text{Newton.meter}}$$

$$\text{hence,} \quad \chi er \in_o = \left.\frac{\text{Farad}}{\text{meter}}\right]$$

The polarisation vector $\vec{P}$ is also related to $\vec{E}$ by equation (4.40), *i.e.,*

$$\vec{P} = \in_o (K-1)\vec{E}$$

Hence by comparison of equation (4.40) and (4.43), we can write

$$\chi er = K - 1$$

The dielectric susceptibility is measure of the capability of the dielectric to take polarisation, by equation (4.45)

$$K = 1 + \chi_{er} \quad ...(4.46)$$

Let us now rewrite equation (4.38),

$$\vec{D} = K \in_o \vec{E} = \in \vec{E} \quad ...(4.47)$$

where $\in$ is the permittivity of the medium,

$$\in = K \in_o$$

$$\text{or} \quad K = \frac{\in}{\in_0} \quad ...(4.48)$$

K is relative permittivity or dielectric constant (sometimes relative permittivity is denoted by $\in_r$) *when a dielectric material is placed in electric field, the distribution of field changes to a degree depending upon relative permittivity.* The χ_{er} is given by using equation (4.43),

$$\chi_{er} = \frac{P}{\in_0 E}$$

$$= \frac{\frac{Q'}{A}}{\frac{\epsilon_0 Q}{K\epsilon_0 A}}$$ (Since $E \frac{Q}{K\epsilon_0 A}$ see equation (4.21) and $P = Q/A$ by definition given in equation (4.27))

$$= K\frac{Q'/A}{Q/A}$$

$$= \frac{\frac{\epsilon Q'}{A}}{\frac{\epsilon_0 Q}{A}} = \frac{\text{Charge density in dielectric}}{\text{Charge density in free space}} \quad ...(4.49)$$

Now from equation (4.46),

$$K = 1 + \chi er$$

This relation clearly shows that the relative permittivity of any matter is larger than unity. The value of χ_{er} of any matter is positive so that K should be larger than one and for vacuum $\chi_{er} = 0$ and therefore $K = 1$.

In the case of *non-linear dielectric including ferroelectrics to be discussed later on, there is no proportionality between vectors* $\vec{P}$ and $\vec{E}$ *as discussed here.*

BOUNDARY RELATIONS

(1) Boundary Conditions at the Surface Separating two Substances

Let us examine, how the electric field changes at the boundary between two different media. The electric field change that occurs in going from one medium to another is determined by the two basic ideas in electrostatics, namely,

(i) the first is Gauss's law. *i.e.,* $\oint_S \vec{D}.d\vec{S} = Q$ and

(ii) the second is that an electrostatics field is a conservative field in other words, on work is done in transporting a charge around a closed path in an electrostatics field, *i.e.,*

$$\oint \vec{E}.\,d\vec{l} = 0$$

We shall apply Gauss's law to the cylindrical surface of height h and base area Δs as shown in Fig. 4.14. The cylindrical box is so constructed that it lies half in each medium. Let D_{1n} be the average normal component of displacement vector $\vec{D}$ to the bottom of the box in medium 1 and D_{2n} the average normal component of displacement vector $\vec{D}$ to the face of the box in medium 2. D_{1n} *is an inward normal.* By making the height of the cylinder h approaching zero, the contribution of the curved surface to the flux is taken as zero. Thus by Gauss's law, the total flux,

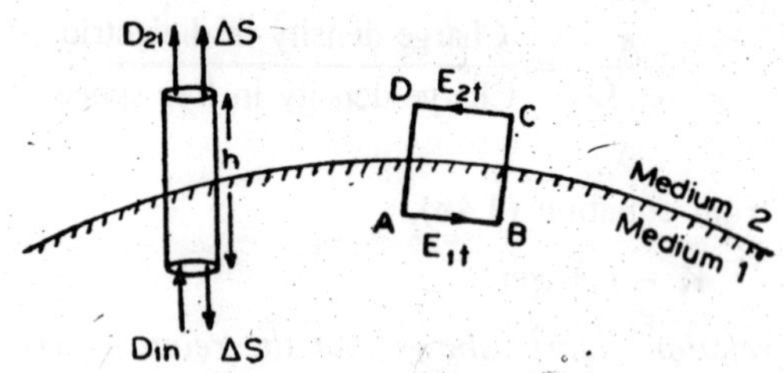

Fig. 4.14

$$D_{2n}\,\Delta S - D_{1n}\,\Delta S = Q$$

where Q is total enclosed by the surface.

The second term of left hand side of this equation is negative because D_{1n} *and* ΔS *are oppositely directed.* We can write,

$$D_{2n} - D_{1n} = \frac{Q}{\Delta S} = \sigma \qquad ...(4.50)$$

where σ is the charge per unit area on the boundary of the two substances. *According to equation (4.50) the normal component of the displacement vector* $\vec{D}$ *changes at a charged boundary between two dielectrics by an amount equal to the surface charge density.*

If the boundary is free from charge $\sigma = 0$, than equation (4.50) reduces,

$$D_{2n} - D_{1n} \qquad ..(4.51)$$

Thus, the normal component of displacement vector is continuous across the charge free boundary between two dielectrics.

We shall now use the idea that the electrostatics field is conservative, the integral of $\vec{E}$ around a closed path,

$$\int_{ABCD} \vec{E}.d\vec{l} = \int_{AB} \vec{E}.d\vec{l} + \int_{BC} \vec{E}.d\vec{l} + \int_{CD} \vec{E}.d\vec{l} + \int_{DA} \vec{E}.d\vec{l} = 0$$

By making the path length BC very small approaching zero, the work along the segments BC and DA of the path normal to the boundary is zero even though a finite electric field may exist normal to the boundary. Therefore, the line integral of $\vec{E}$ around ABCD rectangle is,

$$E_{1t}\,\Delta x - E_{2t}\,\Delta x = 0, \quad AB = CD = \Delta x$$

or
$$E_{1t} = E_{2t} \qquad ...(4.52)$$

Thus the tangential components of the electric field are the same on both sides of a boundary between two dielectrics. In other words, the *tangential electric field is continuous across such a boundary.*

(2) Boundary Conditions at the Surface of a Charged Conductor

If one of the substances (medium or dielectric) considered above is a conductor, for example, if medium 1 is a conductor, then the electric field inside the conductor is zero, consequently, since there are no permanent dipoles in the conductors, the polarisation $\vec{p}$ must be zero, then according to relation ($\vec{D} = \epsilon\,\vec{E} + \vec{p}$) $\vec{D}$ is also zero,

$$D_{1n} = 0$$

Thus equation (4.50) reduces,

$$D_{2n} = \sigma \qquad ...(4.53)$$

This expresses the important result that the normal component of the displacement just out side the conductor is equal to the surface charge density on the conductor. As medium 1 is a conductor ($\sigma_1 \neq 0$) the field E_{1t} in medium 1 must be zero under static condition. Equation (4.52) reduces,

$$E_{2t} = 0 \qquad ...(4.54)$$

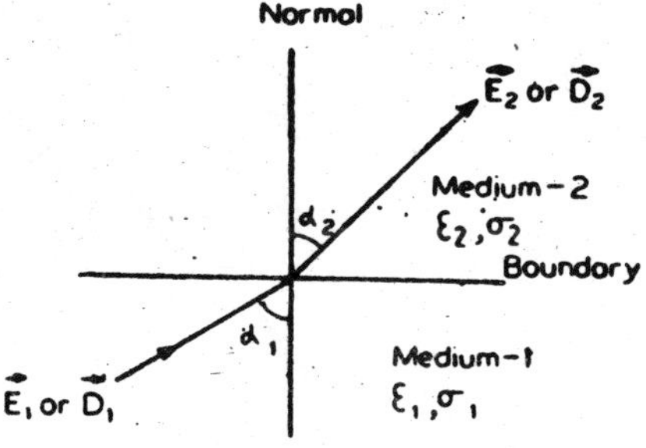

Fig. 4.15

That is, the tangential component of the electric field just outside a conductor (a dielectric conductor boundary) must be zero.

FLUX TUBES

Maxwell and Faraday extended the idea of *flux lines to the flux tube which* have a definite quantitative significance. Let us consider the field intensity at a constant radius from a point charge, the intensity is same for all points on the circumference of this radius.

So that the $\vec{E}$ *and the flux density* $\vec{D}$ *are continuous functions of position around the point charge, and both are constant for a fixed radius.*

Thus, if no flux lines pass through a certain area surrounding the point charge this means flux density is zero there which is not true because $\vec{D}$ and $\vec{E}$ have constant value for all points at equal distances from the charge. It may be that some lines would pass through the area if a larger number of flux lines had been assumed. This difficulty can be avoided by simple extension of the concept of flux lines to flux tubes.

A flux tube is defined as an imaginary tube with walls that are everywhere parallel to $\vec{D}$ *and with a constant electric flux over any cross-section. The requirement that the flux over any cross-section be a constant actually is a consequence of the fact that* $\vec{D}$ *is parallel to the sides of the tube and, therefore, that the flux over the side walls is zero.*

Using the flux-tube idea, we have the flux tubes as shown in Fig. 4.16.

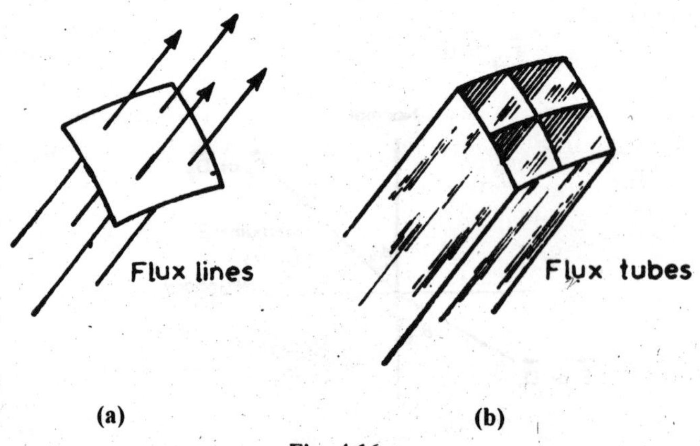

Fig. 4.16

The cross-section of the tube may be of any shape. Conveniently tubes of triangular, squares, or hexagonal cross-section are preferred. In Fig. 4.17 a single flux tube of square cross-section is shown which originates at + Q and ends on –Q.

Fig. 4.17

The magnitude of $\vec{D}$ *varies along the tube but the integral of the normal component of* $\vec{D}$ *over any cross-section of the tube is constant.*

In Fig. 4.17 *the flux density* $\vec{D}$ *is indicated at three points along the tube, where the magnitude of* $\vec{D}$ *is large, the cross-sectional area is small and vice versa. The integral of* $\vec{D}$ *over each cross-sectional area is constant.* All the space can be divided up into tubes of equal flux originating on positive charge.

Number of Tubes

In rationalised RMKS system, we have,

$$\oint_S \vec{D}.\vec{dS} = Q$$

Thus *each unit charge gives rise to a unit tube* (*in place of* 4π *tubes in unrationalised system*).

Faraday Tubes (Unit Tubes of Induction)

In RMKS, each unit charge gives rise to a unit tube irrespective of the surrounding medium. The tubes are then referred as unit tubes of induction. The number or unit tubes of induction due to a charge q placed in vacuum per unit area

$$= \frac{q}{4\pi r^2}$$

$$= \epsilon_0 \frac{q}{4\pi\epsilon_0 r^2} = \epsilon_0 \vec{E} \qquad ...(4.56)$$

i.e., flux density $\vec{D}$.

Maxwell Tubes

However, Maxwell assumed that $1/\in$ unit tubes originate from each unit charge in a medium of permittivity $\in$. Thus the charge q gives rise to $q/\in$ Maxwell's tubes. If we draw a sphere of radius r surrounding the charge, then,

$$\text{No. or Maxwell's tubes per unit area} = \frac{q}{\in 4\pi r^2}$$

$$= \vec{E} \qquad \text{...(4.57)}$$

Hence, the electric intensity at the point is the number of Maxwell unit tubes of force over a unit area surrounding the point.

Maxwell assumed that these tubes are under *longitudinal tension of* $\in_0 \frac{E^2}{2}$ Newton/meter^2 along the direction of the tube which tends to contract and the *lateral pressure of* the same value $\left(\in_0 \frac{E^2}{2} \text{ Newton / meter}^2\right)$ normal to the length, and showed that such a system of tension and pressures in the medium are in equilibrium. He explained that the *attraction between two opposite charges is due to longitudinal tension tending to shorten the tube. The repulsion between the charges could similarly is due to the pressure of the tubes upon each other at right angles to their length and which tend to push the two charges further apart.*

Suppose a tube of force as shown in Fig. 4.18. Apply Gauss's law at P and Q planes of areas dS_1 and dS_2.

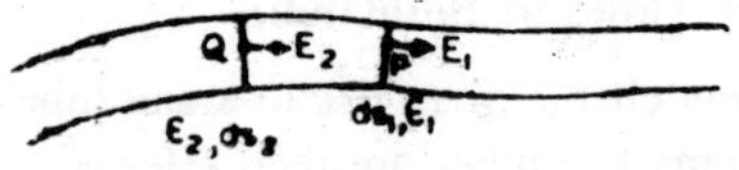

Fig. 4.18

$$\in_1 E_1 dS_1 = \in_2 E_2 dS_2 = 0$$

Since there is no charge enclosed by the tube where $\in_1$ and $\in_2$ are the permittivity constant at these two surfaces.

Thus, $\in_1 E_1 dS_1 = \in_2 E_2 dS_2$

or $$\int_S \vec{D}_1 . d\vec{S}_1 = \int_S \vec{D}_2 . d\vec{S}_2$$

Thus, electric displacement vectors (or flux densities) are different but flux is same.

PHYSICAL MEANING OF POLARISATION OF DIELECTRIC

When a dielectric is placed in an external electric field, the electrically charged particles (atoms, ions, molecules) with their displacement arrange is such order that *dielectric acquires a certain electric moment, such a process is called polarisation.*

Linear dielectrics show a *direct proportionality* between the electric moment P (induced moment) acquired by the particle during the process of polarisation and the intensity of the electric field acting on the particle *i.e.*, the local electric field inside the dielectric E_{int}. Mathematically we represent,

$$P \propto E_{int}$$

$$P = \alpha E_{int} \quad ...(4.58)$$

The coefficient of proportionality α in equation (4.58) is known as polarizability of the given dipole. Polarisability reflects the properties of an individual particle of matter and not of large volume of it hence it is called microscopic electrical parameter of a dielectric.

If, $E_{int} = 1$, in equation (4.58)

then, $$P = \alpha \quad ...(4.59)$$

So the *polarisability* α, *is dipole moment per unit field strength.* If there are N dipoles in unit volume of the matter then equation (4.58) can be written as,

$$P = N \alpha E_{int} \quad ...(4.60)$$

Here P is the average dipole moment of N dipoles

Unit for Polarisability

When dipole moment P is measured in coul. meter, (P = charge × distance) and E is volt/metre, we get SI unit for α, from equation (4.58),

$$\alpha = \frac{\text{Coul.meter}}{\text{Volt / meter}} = \frac{\text{Coul.meter}^2}{\text{Volt}}$$

(coul = volt × Farad)

$$\alpha = \text{Farad. meter}^2$$

MECHANISM OF POLARISATION

Since α is defined in terms of dipole moment, its magnitude will clearly be a measure of the extent to which electric dipoles are formed by ions, atoms and molecules. The *electric dipoles are formed through a variety of mechanisms,* each one contributing to the value of α, hence α is regarded as total polarisability to be the sum of individual polarisabilities each arising from a particular mechanism.

In general, the molecules or atoms (ions) may be affected in the following ways :

(1) Electronic Polarisability

Centres of gravity of the electrons and protons may change resulting in induced dipole moment *i.e.,*

$$P_e = \alpha_e E_{int} \qquad ...(4.61)$$

P_e has all the characteristics of an assembly of dipoles produced by displacement of electrons which have natural frequency equal to, or higher than, those of visible light. So that α_e is called the *optical or electronic polarisability.*

(2) Molecular Polarisability

Molecules having permanent dipole moment *i.e.,* polar molecules are affected in two possible ways:

(a) Atomic Polarisability

Firstly, the field may cause the atoms to be displaced altering the distance between them and hence changing the dipole moment of the molecule. Such polarisability is called atomic represented by αa.

(b) *Orientational or Dipolar Polarisability*

Secondly, the molecule as a whole may rotate about its axis of symmetry, so that the dipoles align with the field. This is referred as orinentational polarisation or dipolar polarization and polarisability is called orientational or dipolar, denoted by α_d.

(3) Interfacial Polarisability

Such type of mechanism is due to large number of defects in the structure of crystals, such as lattice vacancies, impurity centre, dislocation, etc. Polarisability is denoted by α_i; however, this is generally neglected. Thus we can write the total polarisability α as follows :

$$\alpha = \alpha_e + \alpha_a + \alpha_d + \alpha_i \qquad \text{...(4.62)}$$

LORENTZ LOCAL FIELD

In equation (4.60) *i.e.,* $P = N \alpha E_{int}$, it should be noted that this internal intensity of the field, where molecules exist surrounded on all sides by other polarised molecules of matter is, not equal to the external intensity of the field at a given point of a dielectric which is calculated by relation $E = V/d$. The internal intensity E_{int} on a molecule is a geometrical sum of the external intensity E and the component of the intensity caused by the action of the other polarised molecules of the matter on the molecule in question. The following method suggested by Lorentz can be used to estimate the relation between E and E_{int}.

Let us describe a sphere in a dielectric as shown in Fig. 4.19, containing the molecule considered above in its centre. The molecules present inside dielectric outside of the sphere will build up an intensity component E_1 in the centre of the sphere and those inside the sphere an intensity component E_2. Thus internal intensity will be equal to :

$$\vec{E}_{int} = \vec{E} + \vec{E}_1 + E_2 \qquad \text{...(4.63)}$$

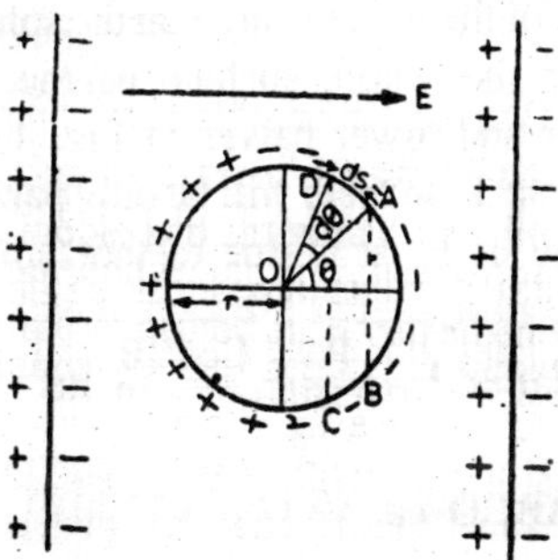

Fig. 4.19

To calculate E_1, let us imagine that the dielectric is removed from the sphere. For the actual pattern of the electric field not to be distorted, a surface electric charge should be placed on the sphere.

At each point of the sphere the surface density of the charge is given by equation (4.31), *i.e.,* $\sigma = \vec{P}.\hat{n}dS$ that is the definition for $\vec{P}$ so that,

$$\sigma = P \cos\theta \quad ...(4.44)$$

where θ is the angle between the direction of $\vec{P}$ *i.e.,* the direction of $\vec{E}$ and the radius of the sphere. The charge dq on the small element of the sphere surface dS will be given by,

dq = σ dS, substituting from Eq. (4.64), we get,

$$dq = P \cos\theta dS \quad ...4.65)$$

The field created by this charge dq inside the sphere,

$$dE_1 = \frac{1}{4\pi \epsilon_o} \frac{dq}{r^2}$$

Putting dq from equation (4.65),

$$dE_1 = \frac{P}{4\pi \epsilon_o r^2} \cos\theta \, dS \quad ...(4.66)$$

The field dE_1 has the component one is parallel to the direction of E, *i.e.,*

$$dE_1 \cos\theta = \frac{P}{4\pi \epsilon_o r^2} \cos^2\theta \, dS \quad ...(4.67)$$

And the other perpendicular to the direction of E *i.e.,*

$$dE_1 \sin\theta = \frac{P}{4\pi \epsilon_o r^2} \cos\theta \sin\theta \, dS \quad ...(4.68)$$

The total intensity E_1 of the entire charge on the sphere is determined by integrating dE_1 over the whole surface of the sphere. But the components of the upper and lower halves in Fig. 4.21 of the sphere perpendicular to E will cancel. So there will be only parallel components. Therefore, the component of E_1 at O due to the right half sphere,

$$= dE_1 \cos\theta = \int_0^{\pi/2} \frac{P}{4\pi \epsilon_o} \frac{\cos^2\theta}{r^2} dS$$

dS is the area or right ABCD *i.e.,*

$$dS = 2\pi r^2 \sin\theta \, d\theta$$

Therefore, $$dE_1 \cos\theta = \int_0^{\pi/2} \frac{P}{4\pi \epsilon_o} \frac{\cos^2\theta}{r^2} 2\pi r^2 \sin\theta \, d\theta$$

$$= \frac{P}{2 \in_0} \int_0^{\pi/2} \text{Cos}^2 \theta \, \text{Sin}\, \theta \, d\theta$$

$$= \frac{P}{2 \in_0} \left[-\frac{\text{Cos}^3\theta}{3} \right]_0^{\pi/2}$$

$$= \frac{P}{6 \in_0} \qquad \text{...(4.69)}$$

As shown in Fig. 4.19 the charges on the surfaces of the two halves of the sphere are equal in magnitude but opposite in sign an equal force acting in the same direction will be exerted on unit positive charge at O. So, the component of E_1 due to left half sphere is also given by,

$$= \frac{P}{6 \in_0}$$

So, the total intensity E_1 is sum of the two i.e,

$$\vec{E} = \frac{\vec{P}}{3 \in_0} \qquad \text{...(4.70)}$$

The direction of this intensity coincides with the direction E, *the value of E_l given in equation (4.70) is called Lorentz field.*

CLAUSIUS MOSOTTI EQUATION

For the simplest case when there is no prevailing effect of any adjacent charges on the molecules in question *i.e.,* $\vec{E}_2$ components of Eq. (4.63) is taken zero, the equation (4.63) after substituting (4.70), gives,

$$\vec{E}_{int} = \vec{E} + \frac{\vec{P}}{3 \in_0} \qquad \text{...(4.71)}$$

We have also a relation between $\vec{P}$ and $\vec{E}$ given in equation (4.40), *i.e.,*

$$P = \in_0 (K - 1)\ \vec{E}$$

or
$$\frac{\vec{P}}{\in_0} = (K-1)\ \vec{E}$$

inserting $\frac{\vec{P}}{\in_0}$ in equation (4.71), we get,

$$\vec{E}_{int} = \vec{E} \left(1 + \frac{(K-1)}{3}\right)$$

$$\vec{E}_{int} = \vec{E}\left[\frac{K+2}{3}\right] \quad ...(4.72)$$

The $\vec{E}_{int}$ given in equation (4.72) is called Mosotti field. Since the expression for $\vec{E}_{int}$ contains K, hence it depends on dielectric material in case of vacuum capacitor with K = 1, that the ratio $\frac{K+2}{3} = 1$ and $\vec{E}_{int} = \vec{E}$.

The value of polarisation P as an electric moment of a unit volume of dielectric is given in equation (4.60) *i.e.,*

$$\vec{P} = N\,\alpha\,\vec{E}_{int}$$

Putting the value $\vec{E}_{int}$ from equation (4.72), in the above equation, we have,

$$\vec{P} = N\alpha\,\frac{(K+2)}{3}\,\vec{E}$$

and equate this expression to P given below:

$$\vec{P} = \epsilon_o (K-1)\,\vec{E}, \text{ [From equation (4.40)]},$$

i.e.,
$$(K-1)\,\epsilon_o E = N\,\alpha\,E\,\frac{K+2}{3}$$

or
$$\frac{K-1}{K+2} = \frac{N\alpha}{3\,\epsilon_o} \quad ...(4.73)$$

where $\alpha = \alpha_e + \alpha_a$. Equation (4.73) is *remarkable for the fact that it shows the connection between the parameters K of a dielectric and its parameters α and N.* The equation (4.73) *is known as Clausius Mosotti equation* As $K = \epsilon/\epsilon_o$ the Clausius Mosotti equation in another simple form is,

$$\frac{\epsilon-\epsilon_o}{\epsilon+2\epsilon_o} = \frac{N\alpha}{3\,\epsilon_o} \quad ...(4.74)$$

The magnitude $\frac{\epsilon-\epsilon_o}{\epsilon+2\epsilon_o}$ or $\frac{K-1}{K+2}$ is known as *specific polarisation of a dielectric.* This is dimensionless because K is dimensionless. It is easy to see that the R.H.S. of equation (4.73) is also dimensionless because N is measured in meter^{-3}, α in farad meter^2 and $\in_0$ in farad/meter, hence $N\alpha/3\in_0$ is dimensionless.

Now using relation $K = 1 + \chi_{er}$. *The Clausius Mosotti equation reduces in terms of the susceptibility of the dielectric.*

$$\frac{\chi_{er}}{\chi_{er}+3} = \frac{N\alpha}{3\in_0} \quad ...(4.75)$$

A version of Clausius Mosotti equation convenient for practical calculation is obtained by multiplying equation (4.73) on both sides by the ratio of the molecular mass M of the substance to its density D.

$$\frac{K-1}{K+2}\frac{M}{D} = \frac{N\alpha}{3\in_0}\frac{M}{D}$$

Since $\frac{NM}{D}$ = Avogadro number N_a

$$\frac{K-1}{K+2}\frac{M}{D} = \frac{1}{3\in_0}N_a\alpha \quad ...(4.76)$$

but $\frac{K-1}{K+2}$ = specific polarisation (P_{sp})

Hence, equation (4.76) reduces,

$$\frac{K-1}{K+2}\frac{M}{D} = P_{sp}\frac{M}{D} \quad ...(4.77)$$

The magnitude given in equation (4.77) is called the *molar polarisation represented by* π,

$$\pi = \frac{K-1}{K+2}\frac{M}{D} = P_{sp}\frac{M}{D} = \frac{1}{3\in_0}N_a\alpha \quad ...(4.78)$$

The dimension of molar polarisation is equal to that of ratio of volume to mass, *i.e.*, the dimension inverse to that of density. For neutral dielectrics the value of K η^2 where η is the refractive index of the medium then we get after substituting in equ. (4.77) and (4.78),

$$\frac{\eta^2-1}{\eta^2+2} = \frac{N\alpha}{3\in_0} \quad ...(4.79)$$

and $$\frac{\eta^2-1}{\eta^2+2}\frac{M}{D} = \frac{N_a\alpha}{3\in_0} \quad ...(4.80)$$

Equations (4.79) and (4.80) are referred *as Lorentz-equation.*

Limitations of the Clausius Mosotti Equation

This equation is derived by taking $E_2 = 0$ (See equation (4.63) for details of E_2 because of the following assumptions:

1. Since polarisation is considered as proportional to the field, it means the polarisation of the molecules by elastic displacement only.
2. Absence of short-range interaction.
3. Isotropy of the polarisability of the molecules.

All these conditions are satisfied with neutral molecules having no constant dipoles *i.e., non polar* (molecules of symmetrical structure). Thus the equation is applicable to *neutral liquids and specially to gases in which the molecules are far apart.*

If a graph is plotted between $\frac{K-1}{K+2}\frac{1}{D}$ on one axis and $\eta = \sqrt{K}$ on the other axis with different values of D, for D = 1, gives the function $\frac{K-1}{K+2}$ (K) approximating unity when K infinitely increases. *i.e.,*

$$\frac{K-1}{K+2} \cong 1 \qquad ...(4.81)$$

Since K in gases is close to unity, we can write,

$$\frac{K-1}{K+2} \cong \frac{K-1}{3}$$

Equation (4.73) reduces,

$$\frac{K-1}{K+2} \cong \frac{K-1}{3} = \frac{N\alpha}{3\,\epsilon_0}$$

or

$$K = 1 + \frac{N\alpha}{3\,\epsilon_0} \qquad ...(4.82)$$

DEBYE EQUATION

When a *polar molecule* is placed in the electric field, it is affected in two ways. *Firstly,* it displaces the centre of gravity of protons and electron, so that an extra dipole moment is induced, giving the *electronic* (atomic) polarisability. For which the *Clausins Mosotti equation is valid. Secondly, the dipole tend to orient* itself so that its potential energy is minimum *i.e.,* it will *tend to align its moment* against the field. The process is called *orientational polarisations. Debye transformed the*

Clausius-Mosotti equation for polar molecules by replacing the electronic and ionic (or atomic) polarisability α of the molecule which is taken in Clausius-Mosotti equation with the sum of deformational polarisability α and the polarisability due to dipole polarisation α_d.

The Clausius-Mosotti equation has no indication of a *temperature variation of permittivity*. However, *for polar dielectrics the polarisability is inversely proportional to temperature due to thermal agitation.* So that orientational polarisation mechanism is temperature dependent. In fact the orientative action of the field is hindered by the thermal agitation. In consequence, a statistical equilibrium is set up in which a slight excess of molecules have their permanent dipoles antiparallel to the field, *i.e.,* the molecules are aligned.

Let us calculate the average component of dipole moment per molecule at temperature T due to alignment. If molecule of permanent dipole moment $\vec{p}_0$ has its axis inclined at an angle θ with the field, then its potential energy is,

$$U = -\vec{P}_0 \cdot \vec{E}, \quad \text{...(4.83)}$$

According to Maxwell Boltzmann law, the number of molecules distributed with axis of their dipoles pointing in all direction within a solid angle d Ω at θ is,

$$dn = A \exp\left[-\frac{U}{K_b T}\right] d\Omega \quad \text{...(4.84)}$$

where K_b is Boltzmann's constant, and A is constant depending on the number of molecules considered.

Integrating this overall possible directions, total number of molecules after putting the value of U from equation (4.83),

$$n = \int_{\theta=0}^{\theta=\pi} A \exp\left[\frac{P_0 E \cos\theta}{K_b T}\right] d\Omega \quad \text{...(4.85)}$$

The effective dipole moment *i.e.,* the dipole moment of a dipole along the field direction = $P_0 \cos\theta$. Therefore, the dipole moment of dn molecules in the field direction,

$$= dn\, P_0 \cos\theta$$

$$= A \exp\left[\frac{P_0 E \cos\theta}{K_b T}\right] P_0 \cos\theta\, d\Omega,$$

Integrating this within limit θ = 0 to θ = π, we get the total dipole moment of all molecules,

$$\sum P_o = \int_{\theta=0}^{\theta=\pi} Ae^{x\cos\theta}\, P_o \cos\theta\, d\Omega$$

where, $x = \dfrac{P_o E}{K_b T}$

The *average dipole moment* ($\overline{P}_o$) along the field,

$$\overline{P}_o = \frac{\sum P_o}{n} = \frac{\int_0^\pi Ae^{x\cos\theta}\, P_o \cos\theta\, d\Omega}{\int_0^\pi Ae^{x\cos\theta}\, d\Omega} \quad ...(4.84)$$

Now the value of dΩ between θ and θ + dθ is calculated by Fig. 4.20

$$d\Omega = \frac{\text{area of ring between } \theta \text{ and } \theta + d\theta}{r^2}$$

$$= \frac{2\pi r \sin\theta\, r d\theta}{r^2}$$

$$= 2\pi \sin\theta\, d\theta \quad ...(4.86a)$$

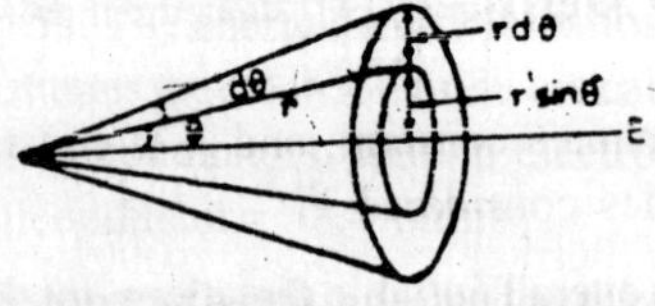

Fig. 4.20

to evaluate the integral of equation (4.86) we put,

Cos θ = α. as θ varies from 0 to π, a will vary + 1 to – 1

Hence. Sin θ dθ = – dα. therefore,

Substituting in equation (4.86), this value we get,

$$\overline{P}_c = \frac{\int_{-1}^{+1} e^{xa} P_o a da}{\int_{-1}^{+1} e^{xa} da}$$

$$\frac{\overline{P}_o}{P_o} = \frac{\left[\frac{a}{x} e^{xa} - \frac{1}{x^2} e^{xa}\right]_{-1}^{+1}}{\left[\frac{1}{x^2} e^{xa}\right]_{-1}^{+1}}$$

$$= \frac{e^x + e^{-x}}{e^x - e^{-x}} - \frac{1}{x}$$

$$= \coth x - \frac{1}{x}$$

therefore, $\frac{\overline{P}_o}{\overline{P}_o} = L(x)$ ...(4.87)

where L (x) = coth x −1/x is called *Langevin function*, because it was derived by Langevin. The relation (4.87) is called *Langevin formula*. The variation of L (x) with x is shown in Fig. 4.21.

The curve is linear for small values of x, *i.e.*, for very small value of E and large values of t, where for large value of x *i.e.*, $\frac{P_o E}{K_b T}$, it approaches unity. Therefore, for *large value of E saturation is obtained i.e., practically all the molecules are aligned with field direction and further increase in E has on effect on average dipole moment.*

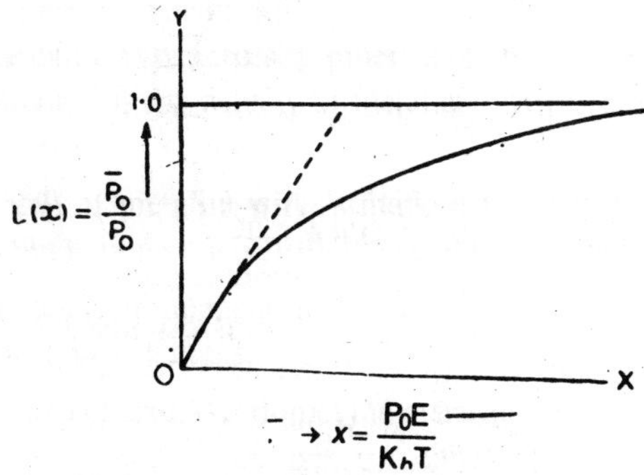

Fig. 4.21

However, the linear region corresponding to small values of x is important at ordinary temperature. The dipole moment P_o of most polar materials is such that x < < 1 for a full range of field strength, even fro those approaching the dielectric strength of the material, so long as the temperature is above about 250° K. Thus a polar dielectric is, in general linear. For small values of x, L (x) can be expanded as,

$$L(x) = \frac{x}{3} - \frac{x^3}{4 \times 5} + \ldots$$

Practically x is very small, hence,

$$L(x) = \frac{x}{3}$$

Hence, equation (4.87) reduces,

$$\frac{\overline{P}_o}{P_o} = \frac{x}{3}$$

$$= \frac{P_o E}{3K_b T}$$

or $$\overline{P}_o = \frac{P_o^2 E}{3K_b T} \qquad \ldots(4.88)$$

E is the polarising field *i.e.*, the field acting on the dipoles. Comparison of this equation (4.88) with equation (4.48), *i.e.*, $P = \alpha E_{int}$ which is for linear dielectric, E_{int} here is E, we get polarisability α_d (*i.e.*, molecular dipole moment per unit polarising field) is,

$$\alpha_d = \frac{P_o^2}{3K_b T} \qquad \ldots(4.89)$$

This represent the orientational polarisability. Therefore, the total polarisability

$$\alpha_{eff} = \alpha + \alpha_d = \alpha + = \frac{P_o^2}{3K_b T} \qquad \ldots(4.90)$$

where α is deformational polarisability as discussed in the beginning ($\alpha = \alpha_e + \alpha_a$). Hence,

$$\alpha_{eff} = \alpha_e + \alpha_a + \frac{P_o^2}{3K_b T}$$

Equation (4.90) is known as the *Langevin-Debye equation.* Therefore, in the case of polar molecule equation (4.73) and (4.78) re given be replacing α by α_{eff} as below:

$$P_{sp} = \frac{\epsilon - \epsilon_0}{\epsilon + 2\epsilon_0} = \frac{K-1}{K+2} = \frac{N}{3\,\epsilon_0}\left(\alpha + \frac{P_o^{\,2}}{3K_bT}\right) \qquad ...(4.91)$$

and
$$\pi = \frac{\epsilon - \epsilon_0}{\epsilon + 2\epsilon_0}\frac{M}{D} = \frac{K-1}{K+2}\,\frac{M}{D}$$

$$= \frac{N_\alpha}{3\,\epsilon_0}\left(\alpha + \frac{P_o^{\,2}}{3K_bT}\right) \qquad ...(4.92)$$

TEMPERATURE DEPENDENCE

Equation (4.90) shows that α_{eff} depends on temp T. Therefore *for polar molecule the polarisability is inversely proportional to the temperature so we can say that K and permittivity also depends on temperature in the same manner as α.* (See Eqn. (4.92)).

Since $\frac{P_o^{\,2}}{3K_bT} >> \alpha$, αvδ for *this reason the permittivity of polar dielectric may be much higher than that of non-polar.* The temperature dependence of dielectric constant K is shown in Fig. 4.22. The intercept of the straight line thus obtained on the y-axis is $\frac{N_\alpha}{3\,\epsilon_0}$ and it slope is $\frac{N_\alpha P_o^{\,2}}{9\,\epsilon_0\,K_b}$

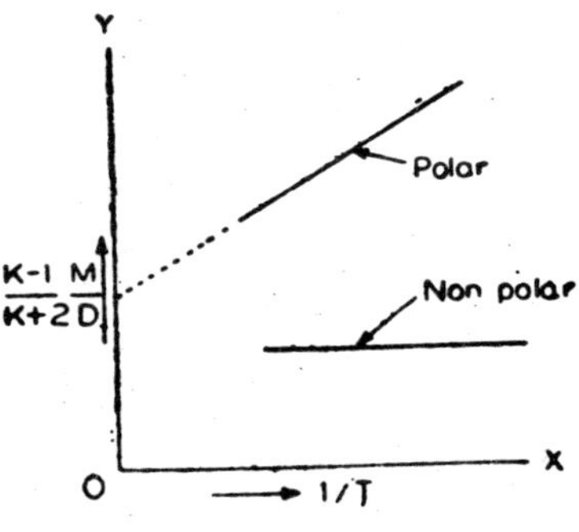

Fig. 4.22

CLASSIFICATION OF DIELECTRICS (FERROELECTRIC AND PARAELECTRIC DIELECTRIC)

The earlier classification of dielectrics into *polar and non-polar is not sufficient.* They *are classified according to their behaviour in electric field.* Take the equation $\vec{P} = \epsilon_0 (k-1)\vec{E}, \vec{E}$ is the resultant of polarising (external field) E_o and field due to polarisation charges E', *i.e.,* $\vec{E}_o + \vec{E}'$ and $\vec{E}$ is related to E_o by relation,

$$\vec{E} = \vec{E}_o / K$$

In most cases the polarisation is proportional to $\vec{E}$.

When polarising field $\vec{E}_o$ is removed then $\vec{E}$ vanishes and in turn $\vec{P}$ also vanishes. But under certain conditions given by equation (4.71),

$$E_{int} = E + \frac{P}{3\epsilon_o}$$

Now when external field is removed, E is zero, then

$E_{int} = \dfrac{P}{3\epsilon_o}$ [P_o is polarisation without external field called permanent polarisation]

Ferroelectric

So we can say that *certain materials, retain a permanent polarisation even though the polarising field is removed, such materials are called ferro-electric by analogy with ferromagnetism.* In other words if a polarisation P_o exists, it will create an electric field at the molecule which tends to polarise the molecules. Such polarisation is called *spontaneous polarisation.* Such materials are able to develop large polarisation when placed in an electric field K >> 1 and $\chi er > 1$.

Examples : Rochelle salt, $BaTiO_3$ *(Barium Titenate),* $SrTiO_3$ *(Stroncium titenate),* $KNbO_3$ *(Potensium niobate).*

Paraelectric

Materials with *small positive susceptibility* due to polarisation in an electric field are termed as para-electric. K and χ_{er} are positive but having small value than ferroelectric. In fact all dielectrics are para with the exception of vacuum, which has precisely zero susceptibility χ_{er} = K – 1 , for vacuum K = 1.

Dia-electric do not Exist

There is no material which has *negative susceptibility.* There is no *analogy with diamagnetism.* In other words, the polarisation in the direction opposite to the polarising field is not possible.

HYSTERESIS LOOP FOR FERROELECTRIC

Fig. 4.23 shows complete curve of *polarisation us, electric field.* As it is the property of ferro-electric to retain polarisation when polarising field is removed, a curve such as in Fig. 4.23 is called *hysteresis loop. Hysteresis means to lag behind,* and it is clear that $\vec{p}$ *lags the electric field vector* $\vec{E}$ when $\vec{E} = 0$, the retained polarisation is ob, it is *spontaneous polarisation (analogous to residual magnetism in magnetisation).* And point C is the electric field which is needed to *completely reverse the polarisation, it is analogous to coercive force in magnetisation.*

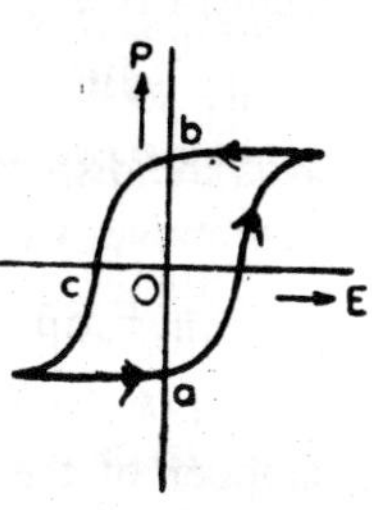

Fig. 4.23

Curie Temperature

However, the *spontaneous polarisation exists only below the certain critical temperature called Curie temperature. Above the* Curie temperature, the *spontaneous polarisation disappears and* substance becomes *para electric.* Below the critical temperature, the variation of susceptibility with temperature is governed by Curie law and above critical temperature by curie Weiss law.

So, we can say that polarisation study of a dielectric is more or less analogous to magnetisation: The behaviour of materials in electric and magnetic field are identical.

ENERGY AND FORCE RELATIONS FOR ELECTROSTATICS FIELD WITH DIELECTRIC PRESENT

The basic energy and pressure laws are modified with dielectrics. When a dielectric is placed in an external field, it is polarised. Let the induced charge + q appears on one surface and – q on the opposite surface, + q and – q are separated by distance d is the case of parallel plate capacitor with dielectric *i.e.,* the thickness of the dielectric slab,

Dipole moment $\vec{p} = qd$. Suppose that while the field $\vec{E}$ is present the charges are further displaced by a distance Δd. The change in dipole moment is given by,

$$dp = q\,(d + \Delta d) - qd$$

or
$$dp = q\,\Delta\,d.$$

The force on the polarisation charge is qE. So the work done for displacement Δ d is given by,

$$dW = q\,E\,\Delta\,d$$

Δd is the displacement in the direction of $\vec{E}$ inserting the value of dP given above, we get,

$$dW = \vec{E}.d\vec{p} \text{ as } \vec{E} \text{ and } \vec{P} \text{ point in the same direction.}$$

This work done is stored in the form o of energy in the dielectric. The supplier of the work is the source of the field, the battery that maintained constant potential difference, between the plates. This stored energy in the dielectric increases the internal energy of the molecules.

But since $\vec{P} = \epsilon_0 (K - 1)\, \vec{E}$, for linear dielectric, where $\vec{P}$ is dipole moment per unit volume. Therefore,

$$d\vec{P} = \hat{I}_0 (K - 1)\, d\vec{E}$$

where $d\vec{P}$ is the change in dipole moment per unit volume. So work done is also given by,

$$\vec{E}.d\vec{P} = \vec{E}.\epsilon_0 (K - 1)\, d\vec{E}$$

Since, $\vec{E}.d\vec{E} = EdE = 1/2\ d\,(E^2)$, therefore,

$$\vec{E}.d\vec{P} = \frac{\epsilon_0}{2}(K - 1)\, d\,(E^2)$$

or $$\int_0^P \vec{E}.d\vec{P} = \frac{\epsilon_0}{2}(K - 1)\, E^2$$

This equation gives the energy change due to polarisation P. Therefore the *total energy density equal to energy density in free space plus the energy density in dielectric.* Therefore, the total energy density,

$$= \frac{1}{2}\epsilon_0\, E^2 + \frac{\epsilon_0}{2}(k - 1)\, E^2$$

$$= \frac{1}{2}\epsilon_0\, KE^2 \qquad ...(4.93)$$

That is *energy increases by K times than that of without dielectric.* Energy per unit volume represented by U is given by,

$$U = \frac{\epsilon_0}{2} KE^2 \text{ joules/meter3}$$

Since $\vec{D} = K\,\epsilon_0\,\vec{E}$ and $\vec{D}$ is in the direction of $\vec{E}$, therefore we can write,

$$U = \frac{1}{2}\,\vec{E}.\vec{D} \text{ joules/meter}^3 \qquad ...(4.94)$$

For any volume,

$$U = \frac{1}{2}\,\epsilon_0 \int_v KE^2 dv \qquad ...(4.95)$$

or
$$U = \frac{1}{2}\int_v \vec{E}.\vec{D}\ dv \qquad ...(4.96)$$

If *dielectric is isotropic* then,

$$U = \frac{1}{2}\,\epsilon_0 K \int_v E^2\ dv \text{ joules} \qquad ...(4.97)$$

Though we have *derived it for parallel plate capacitor but it is applicable in all cases.*

Force

An expression for force can be obtained by *principle of virtual work.* If the work is done by an external agent against the force of attraction between the plates in moving them further apart through a distance Δd,

$$W = \vec{F}.\Delta\vec{d}$$

or
$$W = -F\ \Delta\ d$$

The change in volume due to this additional separation. $\Delta d = A\ \Delta d$, where A is the area of the plate, change in energy due to this change in volume equals $\frac{1}{2}\ \epsilon_0\ KE^2\ A\ \Delta d$

work done = change in energy

$$\frac{1}{2}\ \epsilon_0\ KE^2\ A\ \Delta d = \vec{F}.\Delta\vec{d}$$

$$\frac{1}{2}(\vec{E}.\vec{D})A\ \Delta d = \vec{F}.\Delta\vec{d}$$

Since $\vec{D} = \epsilon_0\ K\ \vec{E}$ and in the direction of $\vec{E}$

or
$$F = \frac{1}{2}\vec{E}.\vec{D}\ A \text{ Newtons}$$

(Since , $\vec{E}$ = Newtons/coul, $\vec{D}$ = Couls/meter^2 and A = meter^2

$$\text{Pressure} = = \frac{\vec{F}}{A} = \frac{1}{2}\vec{E}.\vec{D} \text{ Newtons/meter}^2 \qquad ...(4.98)$$

This is force of attraction between the plates tending to pull them together. Though the pressure exists at any point in an electric field.

Pressure at a Conductor (Plates)—Dielectric Boundary

The pressure is in the direction to draw the conductor into the dielectric. The result follows because the two plates attract each other. The pressure of the conductor on the dielectric is given by equation (4.98) which may be written in several forms,

$$P = \frac{1}{2}\ \vec{E}.\vec{D}\ \frac{1}{2}\frac{D^2}{\epsilon} \text{ since } \vec{D} = \epsilon_0\ K\ \vec{E} \text{ and } K = \frac{\epsilon_0}{\epsilon}$$

$$= \frac{1}{2}\ \epsilon E^2$$

$$= \frac{1}{2}\ \frac{\sigma^2}{\epsilon} \qquad ..(4.99)$$

Since σ is surface density of charge and given by $\sigma = K\ \epsilon_0\ \ E = D$

Alternate Method

As we discussed in Chapter 5, that the potential energy due to a charge distribution is given by,

$$U = \frac{1}{2}\int_v \rho\ V\ dv$$

where v is the volume.

Now according to equation (4.42),

$$\vec{\nabla}.\vec{D} = \rho$$

Therefore, $$U = \frac{1}{2}\int_v V\ (\nabla.\vec{D})\ dv \qquad ...(4.100)$$

but, $$\nabla.(V.\vec{D}) = V\ (\nabla.\vec{D}) + \vec{D}.\nabla V,$$

Substituting in above equation (4.100), we get,

$$U = \frac{1}{2}\int_v \nabla.(V.\vec{D})\ dv\ - \frac{1}{2}\int_v \vec{D}\ \text{grad}\ V\ dv,$$

(Since $(\nabla\ V)$ = grade V). But according to divergence theorem,

$$\int_v \nabla.(V.\vec{D})\ dv = \int_s V\vec{D}.d\vec{S}$$

therefore, $U = \frac{1}{2}\int_S V\vec{D}.d\vec{S} - \frac{1}{2}\int_V \vec{D}.\text{gard } V \, dv$

As we know that $V \; \alpha \; \frac{1}{r}$

$$D \alpha \frac{1}{r^2}$$

$$dS \; \alpha \; r^2$$

So, $\vec{V}.D \; dS \; \alpha \; \frac{1}{r}$

Hence $\int_S VD \, dS$ *will vanish if the ;surface be expanded infinitely, therefore,*

$$U = \frac{1}{2}\int_V \vec{D}.\vec{E}\,dv, \quad (\text{as } E = - \text{ grad } V) \qquad ...(4.101)$$

same as given in equation (4.96).

POTENTIAL AND FIELD DUE TO A POLARISED DIELECTRIC

When a dielectric is polarised the polarisation charge appears.

The polarisation vector $\vec{P}$ is defined by the equation,

$$\vec{P} = \vec{D} - \epsilon_0 \vec{E}$$

Let us take the divergence of this equation. We have,

$$\vec{\nabla}.\vec{P} = \vec{\nabla}.\vec{D} - \epsilon_0 \vec{\nabla}.\vec{E}$$

or $\epsilon_0 \vec{\nabla}.\vec{E} = \vec{\nabla}.\vec{D} - \vec{\nabla}.\vec{P}$

or $\vec{\nabla}.\vec{E} = \frac{1}{\epsilon_0}(\rho f - \vec{\nabla}.\vec{P})$, since $\vec{\nabla}.\vec{D} = \rho f$ by eq. ...(4.42)

As we know that electrostatic potential due to charge distribution ρ is given by,

$$V = \frac{1}{4\pi \epsilon_0}\int_V \frac{\rho}{r} dv',$$

where r = distance from element containing charge to point at which V is to be calculated, and ρ = true charge volume density (couls/meter^3)

When dielectric is polarised the polarisation charge appears and volume charge density due to polarisation is given by equation (4.32a), *i.e.,*

$$\rho_p = -\vec{\nabla}'\vec{P}.$$

Here prime over ∇ *is put to avoid the ambiguity with source point and point of observation because divergence is w.r.t. source point i.e., charge distribution.*

Thus, the resultant volume charge density is

$$\rho - \vec{\nabla}'.\vec{P}$$

Therefore, the total potential due to a polarised dielectric is given by,

$$V_T = \frac{1}{4\pi\epsilon_o}\int_v \frac{\rho - \vec{\nabla}'.\vec{P}}{r}dv'$$

$$= \frac{1}{4\pi\epsilon_o}\int_v \frac{\rho dv'}{r} - \frac{1}{4\pi\epsilon_o}\int_v \frac{\vec{\nabla}'.\vec{P}}{r}dv' \quad \text{...(4.102)}$$

Hence, the *potential* V_p *due to polarisation distribution is given by,*

$$V_p = -\frac{1}{4\pi\epsilon_o}\int_v \frac{(\vec{\nabla}'.\vec{P})}{r}dv' \quad \text{...(4.103)}$$

and the potential due to true charge

$$= \frac{1}{4\pi\epsilon_o}\int \frac{\rho}{r}dv' \quad \text{...(4.104)}$$

where, P = polarisation (couls/meter^2)

r = distance from element containing charge or polarisation to point at which V_T is to be calculated.

The volume integration is taken over all regions containing charge or polarisation.

Dipole Approximation to Polarised Dielectrics

The potential of a dipole is given by,

$$V(r) = \frac{1}{4\pi\epsilon_o}\frac{\vec{p}.\hat{r}}{r^2}$$

or

$$V(r) = \frac{1}{4\pi\epsilon_o}\left[-\vec{p}.\vec{\nabla}\left(\frac{1}{r}\right)\right], \left[\text{since } \nabla\left(\frac{1}{r}\right) = -\frac{1}{r^2}\right]$$

where $\vec{p}$ is the dipole moment.

If there are n dipoles per unit volume, then potential due to unit volume,

$$V(r) = -\frac{1}{4\pi \epsilon_0} n\vec{p}.\vec{\nabla}\left(\frac{1}{r}\right)$$

$$= -\frac{1}{4\pi \epsilon_0} \vec{p}.\vec{\nabla}\left(\frac{1}{r}\right) \quad ...(4.105)$$

where $\vec{p} = n\vec{p}$ *is dipole moment per unit volume. Let us compare equation (4.103) and (4.105),* then we can *interpret the integrand in equation (4.103) as the potential produced by fictitious polarisation by the formation of dipoles over the volume dv'. Thus, the potential* V_p *can then be regarded as the total potential produced by all these dipoles spread through the volume of dielectric. This means that the polarised dielectric is equivalent to an assembly of dipoles in an electric field.*

Now in equation (4.103), the divergence operation (*i.e.*, $\nabla.\vec{p}$) is with respect to the charge distribution *i.e.*, source point where as the potential is determined at any external point say point of observation. Therefore, *we introduce the operator* ∇' *operating on the source coordinates.* The prime on the ∇ is only necessary in integral expressions which relate a field quantity to an integral over a source quantity.

Let (x, y, z) and (x', y', z') be the coordinates of the observation and source points respectively. And the scalar U is,

$$U = U(r)$$

where,

$$r = [(x - x')^2 + (y - y')^2 + (z - z')^2]^{1/2}$$

then,

$$\nabla U = \frac{\delta U}{\delta x}\hat{i} + \frac{\delta U}{\delta y}\hat{j} + \frac{\delta U}{\delta z}\hat{k} \quad ...(4.106)$$

and

$$\nabla' U = \frac{\delta U}{\delta x'}\hat{i} + \frac{\delta U}{\delta y'}\hat{j} + \frac{\delta U}{\delta z'}\hat{k} \quad ...(4.107)$$

But,

$$\frac{\delta U}{\delta x} = \frac{\delta U}{\delta r}.\frac{\delta r}{\delta x}$$

and,

$$\frac{\delta U}{\delta x'} = \frac{\delta U}{\delta r}.\frac{\delta r}{\delta x'}$$

But,

$$\frac{\delta r}{\delta x} = \frac{1}{2}[(x - x')^2 + (y - y')^2 + (z - z')^2]^{-1/2}.\ 2(x - x')$$

$$= \frac{x - x'}{r}$$

and, $\quad \frac{\delta r}{\delta x'} = \frac{1}{2}[(x - x')^2 + (y-y')^2+(z - z')^2]^{-1/2} \times (-2)\ (x-x')$

$$= -\frac{x - x'}{r}$$

therefore, $\quad \frac{\delta r}{\delta x'} = -\frac{\delta r}{\delta x}$

Hence, $\quad \frac{\delta U}{\delta x'} = -\frac{\delta U}{\delta x}$

Similarly, $\quad \frac{\delta U}{\delta y'} = -\frac{\delta U}{\delta y}$ and $\frac{\delta U}{\delta z'} = -\frac{\delta U}{\delta z}$.

putting these values in equation (4.107), we have,

$$\nabla\ U\ (r) = -\ \nabla'\ U\ (r) \qquad ...(4.108)$$

Therefore the ∇ *in equation (4.105) which is with respect to the observation point using equation (4.108), becomes,*

$$V(r) = \frac{1}{4\pi\ \epsilon_0} \vec{P}.\vec{\nabla}'\left(\frac{1}{r}\right)$$

Now $\vec{\nabla}'$ is with respect to source point. *The above relation can be changed into a form that is physically more meaningful by means of Gauss' divergence theorem* and the vector relation.

$$\vec{\nabla}^1.\left(\frac{\vec{P}}{r}\right) = \frac{1}{r}\vec{\nabla}'\vec{P} + \vec{P}.\vec{\nabla}'\left(\frac{1}{r}\right)$$

Substituting $\vec{\nabla}\left(\frac{1}{r}\right)$ in the expression for V(r), we get,

$$V(r) = \frac{1}{4\pi\ \epsilon_0}\left[\vec{\nabla}^1.\left(\frac{\vec{P}}{r}\right) - \frac{1}{r}\vec{\nabla}'.\vec{P}\right]$$

The potential due to the volume distribution of dipoles is found by integrating the expression over the volume which is potential due to polarised dielectrics,

$$V_p = \frac{1}{4\pi\ \epsilon_0}\int\left[\vec{\nabla}'.\left(\frac{\vec{P}}{r}\right) - \frac{1}{r}\vec{\nabla}'.\vec{P}\right] dv'$$

or $$V_p = \frac{1}{4\pi \in_0}\left[\int \vec{\nabla}'.\left(\frac{\vec{P}}{r}\right)dv' - \int \frac{\vec{\nabla}'.\vec{P}}{r}dv'\right]$$

The first term by Gauss' divergence theorem,

$$\int_v \vec{\nabla}'\left(\frac{P}{r}\right)dv = \int_S \frac{\vec{P}.\hat{n}}{r}dS'$$

Thus, $$V_p = \frac{1}{4\pi \in_0}\int_S \frac{\vec{P}.\hat{n}}{r}dS' - \frac{1}{4\pi \in_0}\int \frac{\vec{\nabla}.'P}{r}dv'$$

We know the $\sigma_p = \vec{P}.\hat{n}$ [See equation (4.31)] and $\rho_p = -\vec{\nabla}.'P$ [See equation (4.32a)].

This expression can be interpreted as follows. *The first term, a surface integral, is a potential equivalent to that of a surface charged density while the second term is a potential equivalent to that of a volume charge density. Thus,*

$$V_p = \frac{1}{4\pi \in_0}\int_S \frac{\sigma_p dS'}{r} + \int \frac{\rho_p dv'}{r}, \qquad ...(4.109)$$

Therefore, the total potential produced by a charge distribution ρ in the presence of dielectric or the potential due to a polarised dielectrics given by eq. (4.102) is,

$$V_T = \frac{1}{4\pi \in_0}\int \frac{\rho}{r}dv' + \frac{1}{4\pi \in_0}\int \frac{\rho_p dv'}{r} + \frac{1}{4\pi \in_0}\int_S \frac{\sigma_p dv'}{r} \qquad ...(4.110)$$

The electric field E produced by ρ in the presence of a dielectric can be expressed by,

$$\vec{E} = -\vec{\nabla}\, Vr$$

$$= -\frac{\delta Vr}{\delta r}$$

$$\vec{E} = \frac{1}{4\pi \in_0}\int \frac{\rho dv'}{r^2}\hat{r} + \int \frac{\rho_p dv'}{r^2}\hat{r} + \int \frac{\sigma_p dS'}{r^2}\hat{r} \qquad ...(4.111)$$

where $\hat{r}$ is unit vector directed from the source point to point of observation. The electrostatics field lines begin and end on electric charges. According to Eq. (4.111), the sources of the electrostatics field $\vec{E}$ in the presence of dielectrics are not only the real charges ρ, but also the fictitious, polarisation charges ρ_p and σ_p. Therefore, the field lines of $\vec{E}$ begin and end not only at the points where ρ is present, but also no dielectric surfaces (σ_p) as well as at the points within the dielectrics (r_p). The field lines of $\vec{D}$ on the other hand, begin and end only on the real charge ρ. This follows from the divergence law $\nabla.\vec{D} = \rho$, which shows that $\vec{D}$ is always associated with the real charges only, so that real charges are sources of $\vec{D}$, therefore an $\vec{E}$ and a $\vec{D}$ field may have entirely different structure when dielectrics are present.

TYPES OF CONDENSERS

The condensers are also classified according to dielectric material used. The more common are given below :

1. Variable air condensers
2. Condenser with solid dielectric. The solid dielectric commonly used are :
   (a) Mica—then we call mica condenser.
   (b) Paper—paper condenser are made by taking two strips of tin or aluminium foil insulated by paper and rolling it into a bundle. They can be safely used up to 500 volts.
   (c) Ceramic—for large capacitance using ceramic as dielectric.
3. Electrolytic condenser: It is made by putting two aluminium plates in an electrolyte.

DIELECTRIC STRENGTH

The *dielectric material loses its property of insulation for a particular field strength. It is referred as breakdown strength.* Let us examine what occurs in a material when breakdown occurs. When a dielectric is placed in an electric field, a strain is produced on the elastic forces (by which the electrons are bound to the nucleus or the atoms are bound together in a molecule) which permits the polarisation of the molecule. Under normal condition, the electrons are not able to leave the molecule, should the electric field by made so high that electrons can move away from their normal position on the molecule. The electrons will drift to the positive electrode and the positive ion that results when electron is removed will drift to the negative electrode. *Because of this drift of charged carriers, the dielectric is no longer an insulator and dielectric break down is said to have occurred.* The magnitude of the electric field at which the dielectric breakdown occurs in an insulating material is the dielectric strength of the material.

Dielectric strength of some common materials are given below :

Material	*Dielectric strength KV/m*
Air	3,000
Mica	1,00,000
Rubber	40,000
Porcelain	10,000
Transformer Coil	12,000

SOLVED EXAMPLES

Example 1:

Using data of example 4.6 calculate $\vec{E}$, $\vec{D}$ *and* $\vec{P}$ *then verify the equation,*

$$\vec{D} = \epsilon_0 \vec{E} + \vec{P}$$

is correct both in the gap and in the dielectric.

Solution:

In the dielectric

$\vec{D} = K\epsilon_0 \vec{E}$, putting the value of E and K from example 4.6.

$$\vec{D} = 7 \times 8.9 \times 10^{-12} \text{ couls}^2/\text{Newton. meter}^2 \times 1.4 \times 10^3 \text{ volts/meter}$$

$$= 8.9 \times 10^{-8} \text{ couls/meter}^2, \text{ volt} = \text{Newtons. meter}$$

and $\vec{P} = \epsilon_0 (K - 1)\vec{E}$

$$= 8.9 \times 10^{-12} (7 - 1) (1.43 \times 10^3)$$

$$= 7.5 \times 10^{-8} \text{ coul/meter}^2$$

The D_0 in air gap is given by,

$$D_0 = \epsilon_0 E_0 \quad \text{as } K = 1 \text{ for air}$$

$$= 8.9 \times 10^{-12} \text{ K } 1 \times 10^4$$

$$= 8.9 \times 10^{-8} \text{ coul/meter}^2$$

In the air gap K = 1, hence,

$$P_o = \epsilon_o (K - 1)E_o = 0$$

in the dielectric,

$$\vec{D} = \epsilon_0 \vec{E} + \vec{P} = 8.9 \times 10^{-12} \times 1.4 \times 10^2 + 7.5 \times 10^{-8} \text{ coul/meter}^2$$
$$= 8.9 \times 10^{-8} \text{ coul/meter}^2$$

hence verified.

In the air gap $\vec{P} = 0$

$$D_o = \epsilon_o E_o$$
$$= 8.9 \times 10^{-12} \times 1 \times 10^4$$
$$= 8.9 \times 10^{-8} \text{ coul/meter}^2$$

We can say that the relation $\vec{D} = \epsilon_0 \vec{E} + \vec{P}$ holds in general, or

$$\vec{D} = \epsilon_o \vec{E} + \vec{P}, \text{ everywhere}$$

Example 2:

In example 6.7 if we maintain a constant potential difference 100 volts after the dielectric is introduced, calculate the energy.

Solution:

The electrostatics energy without dielectric per unit volume,

$$= \frac{1}{2} \epsilon_0 E_0^2 \text{ jouls/meter}^3$$

and the energy density with dielectric $= \frac{1}{2} K \epsilon_o E_o^2$

substituting, K and E_o, we get,

$$U = \frac{1}{2} \times 7 \times 8.9 \times 10^{-12} \times 1 \times 10^8 \text{ jouls/meter}^3$$
$$= 31.15 \times 10^{-4} \text{ jouls/meter}^3$$

So energy density increases by K times.

(b) *However if source of emf is withdrawn as soon as dielectric is introduced then the examples 4.10 (a) reduces.*

Solution:

Potential difference after dielectric is introduced

$$= V_o / K$$

where V_0 is potential difference before introducing the dielectric,

$$C = KC_0,\ U = \frac{1}{2}\ CV^2 = \frac{1}{2}KC_0\left(\frac{V_0}{K}\right)^2 = \frac{1}{2}\frac{C_0V_0^{\ 2}}{K}$$

$\frac{1}{2}C_0V_0^{\ 2}$ *is the energy before the dielectric is introduced. Hence energy decreases by factor 1/K.*

Example 3:

The polarisability of NH_3 molecule in gaseous state, from the measurement of dielectric constant is found 2.5×10^{-39} farad. meter^2 and 2.0×10^{-39} farad-meter^2 at temperature 300° K and 400° K respectively. Calculate the contribution to the polarisability because of deformation of molecules and the contribution because of permanent dipole at these temperatures.

Solution:

Using equation (4.90) of Langevin theory,

$$\alpha_{eff} = \alpha \text{ (due to deformation)}$$

$$+ \frac{P_0^{\ 2}}{3K_bT} \text{ (due to dipole moment } P_0)$$

On putting the values,

$$2.5 \times 10^{-39} = \alpha + \frac{P_0^{\ 2}}{3K_b.300}$$

or $$2.5 \times 10^{-39} = \alpha + \frac{\beta}{300},\ \text{where } \beta = \frac{P_0^{\ 2}}{3K_b}$$

and $$2 \times 10^{-39} = \alpha + \frac{\beta}{400}$$

Solving these two equations,

$$\beta = 6 \times 10^{-37} \text{ farad.meter}^2$$

hence, $\alpha = 0.5 \times 10^{-39}$ farad.meter^2

The deformational polarisability = 0.5×10^{-39} farad.meter^2 and is independent of temperature.

Due to orientational polarisation at temperature 300°K, the polarisability,

$$= \frac{\beta}{T} = \frac{6 \times 10^{-37}}{300} \text{ Farad.meter}^2$$

$$2 \times 10^{-39} \text{ Farad.meter}^2$$

at temperature 400°K $= \dfrac{6 \times 10^{-37}}{400}$

$$= 1.5 \times 10^{-39} \text{ Farad.meter}^2$$

Example 4:

The plates of parallel plate capacitor are separated by 1.00 mm. What must be the plate area if the capacitance to be 1.0 Farad assuming that air is filled between the plates.

Solution:

Using equation (4.3), we get,

$$A = \frac{dC}{\epsilon_0} = \frac{1.0 \times 10^{-3} \text{ metre (1.00 Farad)}}{8.9 \times 10^{-12} \text{ coul}^2 / \text{newton. metre}^2}$$

as, $\quad \text{Farad} = \dfrac{\text{Coul}}{\text{Volt}} = \dfrac{\text{Coul}^2}{\text{newton.metre}}$

Hence, $\quad A = 1.1 \times 10^8 \text{ metre}^2$

This is approximately the area of a square sheet of more than 9 km of side. Indeed farad is very large unit. Therefore, we use smaller unit as micro farad or micro-micro farad.

Example 5:

A parallel plate capacitor is filled with two dielectrics of same dimensions but different dielectric constant K_1 *and* K_2 *respectively as shown in Fig.* 4.10*a and b. calculate its capacitance.*

Solution:

From Fig. 4.24 (a) the arrangement is simply two capacitors in parallel, and each capacitor will have the area A/2.

Total capacitance $C = C_1 + C_2$

$$= \frac{\epsilon_0 A / 2 k_1}{d} + \frac{\epsilon_0 A / 2 k_2}{d}$$

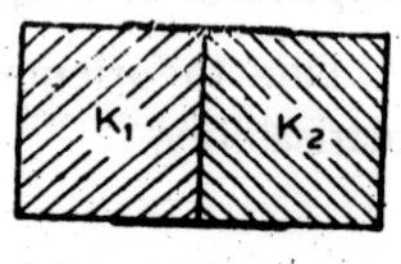

Fig. 4.24(a)

$$= \frac{\epsilon_0 A}{d}\left(\frac{k_1 + k_2}{2}\right)$$

The arrangement of Fig. 4.10b can be regarded as two capacitors in series. Each capacitor will be of area A and separation d/2. Therefore the total capacitance is given by the relation,

Fig. 4.24(b)

$$\frac{1}{C} = \frac{1}{C_1} + \frac{1}{C_2} = \frac{d}{2}\left(\frac{1}{\epsilon_0 AK_1} + \frac{1}{\epsilon_0 AK_2}\right)$$

$$\text{or } C = \frac{2\epsilon_0 A}{d}\left(\frac{K_1 K_2}{K_1 + K_2}\right)$$

Example 6:

The separation between the plates of parallel plate capacitor is d, one plate is at zero potential. The space between the plates is filled with a material whose dielectric constant varies uniformly from one plate to the other. If K_1 *and* K_2 *are the values of the dielectric constant at the two plates, show that the capacity per unit area is*

$$C = \frac{\epsilon_0}{d}\frac{K_2 - K_1}{\log_e \frac{K_2}{K_1}}$$

Solution:

The electric intensity E is given by,

$$E = \frac{\sigma}{\epsilon_0 K}, \text{ since } E = \frac{E_0}{K}$$

As K is not constant, but varies from K_1 to K_2 increase in dielectric constant per unit length $= \frac{K_2 - K_1}{d} = \alpha$. The dielectric constant at point at a distance x from the first plate is $K = K_1 + \alpha x$.

Hence the electric intensity at this point $= \frac{\sigma}{\epsilon_0 (K_1 + \alpha x)}$

Potential difference between the plates,

$$V = -\int_{\alpha}^{0} \vec{E}.d\vec{x} = \int_{0}^{d} E dx$$

$$= \frac{\sigma}{\alpha \epsilon_0} \log \frac{K_1 + \alpha d}{K_1}, \text{ substituting } \alpha$$

$$= \frac{\sigma d}{\epsilon_0 (K_2 - K_1)} \log \frac{K_1 + \frac{(K_2 - K_1)d}{d}}{K_1}$$

$$= \frac{\sigma d}{\epsilon_0 (K_2 - K_1)} \log_e \frac{K_2}{K_1}$$

Hence, $C = \frac{\text{total charge}}{V} = \frac{\sigma A}{V}$

$$= \frac{A \epsilon_o (K_2 - K_1)}{d \log_e K_2 / K_1}$$

Example 7:

A parallel plate capacitor consists of two square metal plates 5.0cm of side and separated by 1 cm. A sulphur slab 6 mm thick is placed on the lower plate, calculate the capacitance of the capacitor, dielectric constant of sulphur is 4.

Solution:

When sulphur slab of 6 mm is placed on the lower plate, it leaves an air gap of 4 mm thickness. The capacitor can be regarded as two capacitors in series.

An air capacitors of 4 mm plate spacing and a sulphur field capacitor of 6 mm plate space as shown in Fig. 4.11. From Equation (4.3) the capacitance of air capacitor,

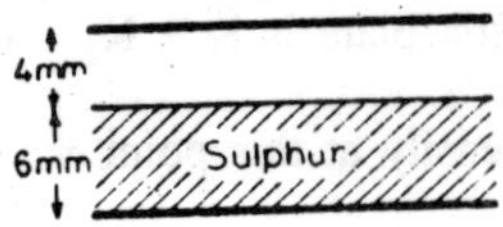

Fig. 4.25

$C_a = \frac{\epsilon_o A}{d}$ farad when A and d are in metres

$$= 8.85 \times 10^{-12} \frac{A}{d} \text{ farad}$$

$$= 8.85 \frac{A}{d} \ \mu\mu F,$$

since $A = 0.5 \times 0.5$
and $d = 4$ mm $= 0.004$ metre

hence, $= \dfrac{8.85 \times 0.5^2}{0.004}$

$$= 553 \ \mu\mu F$$

The capacitance of the sulphur filled capacitor by Equation (4.26),

$$C_s = \frac{K \in_0 A}{d'}$$

$$= \frac{8.85 \times 4 \times 0.5^2}{0.006} \mu\mu F$$

$$= 1,475 \ \mu\mu F$$

Then total capacitance,

$$\frac{1}{C} = \frac{1}{C_a} + \frac{1}{C_s}$$

$$C = \frac{C_a C_s}{C_a + C_s} = 402 \ \mu\mu F$$

Example 8:

A concentric spherical capacitor is formed is formed by two concentric spheres of radii a and b. If the medium between the spherical shells has the dielectric constant K_1 *from a or r and* K_2 *from r to b show that*

$$C = \frac{4\pi \in_0}{\left[\frac{1}{k_1 a} - \frac{1}{k_2 b} + \frac{1}{r}\left(\frac{1}{k_2} - \frac{1}{k_1}\right)\right]}$$

Solution:

See Fig. 4.3. The potential difference between the inner and outer spheres is given by,

$$V_a - V_b = -\int_b^a E dr = +\int_a^b \frac{Q}{4\pi \in_0 Kr^2} dr$$

$$= \frac{Q}{4\pi \in_o} \int_a^b \frac{dr}{Kr^2}$$

AS $K = K_1$ from a to r and $K = K_2$ from r to b,

$$V_a - V_b = \frac{Q}{4\pi \in_o} \left[\int_a^r \frac{dr}{K_1 r^2} + \int_r^b \frac{dr}{K_2 r^2} \right]$$

$$= \frac{Q}{r\pi \in_o} \left[\left| -\frac{1}{K_1 r} \right|_a^r + \left| -\frac{1}{K_2 r} \right|_r^b \right]$$

$$= \frac{Q}{4\pi \in_o} \left[\frac{1}{r} \left(\frac{1}{K_2} - \frac{1}{K_1} \right) + \left(\frac{1}{K_1 a} - \frac{1}{K_2 b} \right) \right]$$

Therefore the capacitance

$$C = \frac{Q}{V_a - V_b}$$

$$= \frac{4\pi \in_o}{\frac{1}{r} \left(\frac{1}{K_2} - \frac{1}{K_1} \right) + \left(\frac{1}{K_1 a} - \frac{1}{K_2 b} \right)}$$

Example 9:

For a parallel plate capacitor following data are given—area of the each plate = 100 cm^2*; separation of plates = 1.00 cm A potential difference of 100 volt is applied with no dielectric present. A dielectric slab of dielectric constant 7.00 and of thickness 0.50 cm is introduced. Calculate:*

(1) the free charge

(2) electric field strength in the gap,

(3) electric field strength in the dielectric,

(4) potential difference between the plates,

(5) capacitance with dielectric.

Solution:

Capacitance before the dielectric is introduced,

$$= C_o = \frac{\in_o A}{d} = 8.9 \times 10^{-12} \times 10^{-2}/10^{-2} = 8.9 \ \mu\mu F$$

(1) Free charge $q = C_o V_o = 8.9 \times 10^{-12}$ Farad $\times$ 100 volt

$= 8.9 \times 10^{-10}$ coul

(2) Electric field strength in the gap. Using the relation,

$$E_0 = \frac{Q}{\epsilon_0 A}$$

$$= \frac{8.9\times10^{-10}}{8.9\times10^{-12}\times10^{-2}} = 1\times10^4 \text{ volt / metre}$$

(3) Electric field in the dielectric,

$$E = \frac{E_0}{K},$$

$$= \frac{1\times10^4}{7} = 0.14\times10^4 \text{ volts/metre}$$

(4) Potential Difference between the plates

$$V = E_o (d - t) + Et$$

Substituting E and E_o from above,

$$V = 1 \times 10^4 (0.5 \times 10^{-2}) + 0.14 \times 10^4 (0.5 \times 10^{-2})$$

$$= 57 \text{ volts}$$

(5) Capacitance with slab

$$\frac{Q}{A} = \frac{8.9\times10^{-12}\text{ coul}}{57 \text{ volt}} = 16\ \mu\mu F$$

Example 10:

Boundary between two dielectrics. Let two isotropic dielectrics media 1 and 2 be separated by a charge free plane boundary. Let permitivities be ϵ_1 and ϵ_2 then show that,

$$\frac{\tan\alpha_1}{\tan\alpha_2} = \frac{K_1}{K_2}$$

where K_1 and K_2 are the dielectric constants of the two medium.

Solution:

As given media 1 and 2 are separated by a charge free plane boundary, *i.e.*, surface density of charge on boundary is zero, then boundary relations are:

$D_{1n} = D_{2n}$ and $E_{1t} = E_{2t}$ as discussed above in Eq (4.51) and (4.52)

From figure,

D_{1n} *i.e.,* normal component of $D_{1n} = D_1 \cos x_1$

and $D_{2n} = D_2 \operatorname{Cos} \alpha_2$

while, $E_{1t} = E_1 \sin \alpha_1$ and $E_{2t} = E_2 \sin \alpha_2$

therefore, $D_1 \cos \alpha_1 = D_2 \cos \alpha_2$, (since $D_{1n} = D_{2n}$)

$E_1 \sin \alpha_1 = E_2 \sin \alpha_2$, (since $E_{1t} = E_{2t}$)

Ratio of these,

$$\frac{D_1 \operatorname{Cos} \alpha_1}{E_1 \operatorname{Sin} \alpha_1} = \frac{D_2 \operatorname{Cos} \alpha_2}{E_2 \operatorname{Sin} \alpha_2}$$

But $D_1 = \epsilon_1 E_1$,

and $D_2 = \epsilon_2 E_2$

$$\frac{\epsilon_1 E_1 \operatorname{Cos} \alpha_1}{E_1 \operatorname{Sin} \alpha_1} = \frac{\epsilon_2 E_2 \operatorname{Cos} \alpha_2}{E_2 \operatorname{Sin} \alpha_2}$$

or

$$\frac{\epsilon_1}{\tan \alpha_1} = \frac{\epsilon_2}{\tan \alpha_2}$$

or

$$\frac{\tan \alpha_1}{\tan \alpha_2} = \frac{\epsilon_1}{\epsilon_2}$$

It K_1 and K_2 are the relative permittivities of two medium, then,

$$\frac{\tan \alpha_1}{\tan \alpha_2} = \frac{K_1 / \epsilon_o}{K_2 / \epsilon_0} = \frac{K_1}{K_2} \qquad ...(4.55)$$

Example 11:

A potential of 50 KV is applied between a pair of parallel plates which are 0.02 m apart the medium between them being air. A sheet of mica of thickness 0.0075 m is introduced. Examine whether the breakdown will occur or not. Given the dielectric constant of mica is 4 times than that of air. The breakdown strength of air is 3000 KV.m and mica 1,00,000 KV/M.

Solution:

See the geometry of parallel plate capacitor. Before the addition of mica $E = \frac{V}{d} = \frac{50}{0.02} = 2500 \text{ KV/m}$. Since the dielectric strength of air is 3,000 KV/m, which is higher than the potential gradient in the field. There will be no breakdown. When mica plate is introduced the potential difference.

$$V = Ed + E_1 d_1$$

But as electric displacement vector is constant,

$$D = \epsilon_0 KE, \text{ everywhere}$$

and $\epsilon_0 KE = \epsilon_0 K_1 E_1$, E_1 is field inside mica.

or $KE = K_1 E_1$

As given $K_1 = 4K$

we get, $E = 4E_1$

or $E_1 = E/4$.

Substituting in the above equation and noting that air space,

$$= 0.02 - 0.0075$$

$$= 0.0125 \text{ metre,}$$

hence equation $V = Ed + E_1 d_1$ after substitution is,

$$50 \text{ KV/m} = 0.0125 \text{ E} + 0.0075 \text{ E/4}$$

Solving for E, we get after substituting E given above

i.e., $E_1 = E/4$,

$$E = 3{,}440 \text{ KV/m.}$$

Since this field strength exceeds the dielectric strength of air, the air will break down, since the air after breakdown is conducting then the field intensity in the mica becomes,

$$E_1 = \frac{50 \text{KV}}{0.0075}$$

$$= 66{,}700 \text{ KV/m.}$$

But the breakdown strength of mica is greater than this value, and mica will not bread-down.

So we cannot introduce the mica because air will breakdown and there will be sparking in the system.

Table 4.1 : Three Electric Vectors

Name	Symbol	Associated with	Boundary condition
Electric field	$\vec{E}$	All charges	Tangential component continuous Eq. (4.52)
Electric displacement or Flux Density	$\vec{D}$	Free charges	Normal component continuous Eq. (4.51)
Polarisation (Electric dipole moment per unit volume)	$\vec{P}$	Polarisation charge only	Vanishes in a vacuum Eq. (4.40)
Defining equation for		$\vec{E}$ $\vec{F} = q\vec{E}$, Eq. (4.10)	
General relation among the three vectors,		$\vec{D}\ \epsilon_0\ \vec{E} + \vec{P}$, Eq. (6.36)	
Gauss' law in presence of Dielectric media		$\oint \vec{D}.d\vec{S} = Q$, (Q is free charge) Eq. (4.41) or $\nabla.\vec{D} = \rho_p$, Eq. (6.42)	
Empirical relations for certain dielectric materials		$\vec{D} = K\epsilon_0\vec{E}$, Eq.(6.38) $\vec{P} = (K-1)\epsilon_0\vec{E}$, Eq.(6.40)	

EXERCISES

1. (a) Let n capacitors of capacitance $c_1, c_2, \ldots c_n$ be connected in series . Show that effective capacitance c is given by,

$$\frac{1}{c} = \sum_{i=1}^{n} \frac{1}{c_i}$$

(b) Let n capacitors of capacitance $c_1, c_2, \ldots c_n$ be connected in parallel. Show that effective capacitance

$$c = \sum_{i=1}^{n} c_i$$

2. The capacitance of a parallel plate capacitor is 400 picofarad and its plates are separated by 2 mm of air. (1) What will be

the energy when it is charged to 1500 volts ? (2) What will be the potential difference with same charge if plate separation is doubled ? (3) How much energy is needed to double the distance between its plates ?

3. A one core lead sheeted cable has conductor core of 0.5 cm diameter and the lead sheet has diameter 1.5 cm. The insulating material is rubber. At what voltage will the insulation break down. Given that rubber has a dielectric strength of 400KV/cm.
4. A co-axial cable consists of a copper core of 1 mm radius within outer metal sheet of 1 cm radius separated by an insulating material of dielectric constant 5. What is the capacitance per metre length of the cable in picofarad?
5. If half the space between two concentric conducting spheres by filled with dielectric of dielectric constant K and the rest is filled with air. Show that capacitance of the capacitor thus formed will be same as if the whole part is filled with the dielectric of dielectric constant 1/2 (1 + K).
6. Calculate the induced dipole moment per unit volume of he gas if placed in a field of 6000 volts/cm. The atomic polarisability of he is 0.21×10^{-24}cm^3 and density of he is 20.6×10^{19}atoms/cc.
7. An atom of oxygen on being polarised produces a dipole moment of 1.5×10^{-23} stat coul × cm. If the distance of the centre of the negative charge could from the nucleus is 4×10^{-15} cm, calculate the polarisability of oxygen atom.
8. Define capacitance and capacitor. On what factors the capacitance depends ?
9. Calculate the capacity of a parallel plate condenser. What will be the capacity if the space between the plates is partially filled with a slab of thickness t and dielectric constant K ?
10. Calculate the capacity of a condenser, consisting of two spheres of radii a and b, separated by a dielectric of S.I.C.K.
11. Find an expression for the capacity per unit length of two coaxial conducting cylinders placed in air.
12. Obtain an expression for energy and force of attraction between parallel plate capacitor.
13. What are polar and non-polar molecules ? Give examples.

14. What do you mean by dielectric polarisation ? What is uniform and non-uniform polarisation ? Show that for non-uniform polarisation

$$\nabla . p = -\rho_p$$

15. Define $\vec{D}$, $\vec{E}$ and P. Establish the relation

(1) $\vec{D} = \epsilon_o \vec{E} \vec{P}$

(2) $\nabla . \vec{D} = \rho_f$

Symbols have usual meaning.

16. Consider a surface separating two dielectric media of permittivity ϵ_1 and ϵ_2. If the dielectric field intensity in medium 1 makes an angle θ_1 with the normal to the surface. Show that the angle θ_2 which the field intensity in median 2 makes with the normal to the surface is given by

$$\epsilon_1 \text{Cos } \theta_1 = \epsilon_2 \text{ Cos } \theta_2$$

17. Define a flux tube. How the force of attraction and repulsion between charges is explained by the idea of Maxwell tubes ?

18. Explain what you mean by electronic, ionic and dipolar polarisabilities.

19. If E denotes the electric intensity in a homogeneous and isotropic dielectric, show that the actual field acting on one of the molecules of the dielectric is

$$\vec{E}_{eff} = \vec{E} + \frac{\vec{P}}{3 \epsilon_0}$$

20. Derive the Classius-Mosotti's equation. How this equation is modified by Debye ?

21. What is meant by polarisability of a molecule ? Give in brief Langevin Debye theory of dielectric polarisation.

22. Derive a relation between electric susceptibility and atomic polarisability.

23. Show that the potential due to a polarised dielectric is the same as the potential due to a surface charge density over the surface of volume which has been polarised plus that due to a volume charge density.

23. What do you mean by dielectric strength ?

Multiple Choice Questions

1. $\nabla.\vec{D} = \rho$ is based on

   (a) Ampere's law (b) Faraday's law

   (c) Ohm's law (d) Gauss' law

2. The effective capacitance is reduced when capacitors are connected in

   (a) series

   (b) parallel

   (c) series-parallel combination

   (d) none of the above.

3. The unit of $\vec{D}$ is

   (a) V.m^2, (b) Coul/m^2,

   (c) V/m (d) Q/m

4. The relation $\vec{D} = \epsilon_0 \vec{E} + \vec{P}$ holds good

   (a) only in vacuum

   (b) only inside dielectric

   (c) only outside dielectric

   (d) everywhere.

5. Poisson's equation states that

   (a) $\nabla^2 V = -\rho/\epsilon$ (b) $\nabla^2 V = \rho/\epsilon$

   (c) $\nabla V = \rho/\epsilon$ (d) $\nabla V = \rho^2/\epsilon$

6. By inserting a plate of a dielectric material between the plates of a parallel plate capacitor, the energy (stored) is increased five times. The dielectric constant of the material is

   (a) 1/25, (b) 1/5,

   (c) 5 (d) 25

7. The unit of polarisation $\vec{p}$ is

   (a) same as that of $\vec{E}$

   (b) same as that of $\vec{D}$

(c) same as that of $\vec{E}$/ coulomb

(d) same as that of charge

8. Which statement influences the capacity of a capacitor?

(a) area of the plates, thickness of the plates and the rate of charge.

(b) area of the plates, dielectric and the rate of charge.

(c) distance between the plates, dielectric and thickness of the plate.

(d) distance between the plates, area of the plates and dielectric.

9. To increase the capacitance of capacitor, the plates must be placed

(a) further apart (b) closer together

(c) in series (d) made smaller.

10. The effect of the dielectric is to

(a) increase the capacitance

(b) decreases the capacitance

(c) reduce the working voltage

(d) increase the distance between the plates,

11. Electric flux density is the ratio of

(a) charge to area of dielectric at right angles to direction of electric flux.

(b) charge to distance between the plates.

(c) electric flux to distance between the plates.

(d) electric flux to are of the plates.

12. Which of the following is correct for a capacitor

(a) $W = \frac{1}{2}\frac{Q_2}{C}$ (b) $W = \frac{1}{2} CI^2$

(c) $W = \frac{1}{2} VC^2$ (D) $W = \frac{1}{2} CV^2$

5

Current Electricity

OHM'S LAW AND RESISTANCE

1. $I \propto V$ or $V = IR$ or $I = GV$, provided physical conditions (temperature, magnetic field, radiation etc.) remain unchanged.

   Where G = conductance = $\frac{1}{R}$. [R = resistance].

   $G = \frac{ne^2\tau}{2m}\left(\frac{A}{l}\right)$, $G = J/E$. Where m = Mass of electron, τ = Relaxation time, E = Intensity of electric field, n = Number of free electrons A = Area, l = length.

   S.I. unit of G = mho = $ohm^{-1} = \frac{1}{ohm} = \frac{amp}{volt}$ siemen.

2. $R = \rho\frac{l}{A}$. where l = length of wire (along the direction of flow of I), A = cross sectional area ($\perp$ to the direction of flow of I) & ρ = specific resistance or electrical resistivity.

   σ (electrical conductivity or specific conductance) = $\frac{1}{\rho}$. Specific conductance $(\sigma) = \frac{J}{E}$

   Where J = Current density, E = Intensity of electric field.

   S.I. unit of ρ & σ = ohm (Ω) × metre & mho/metre respectively.

3. $R = \rho\frac{l}{\pi r^2}$. ($A = \pi r^2$ for wire) $= \rho\frac{l / \pi D^2}{4}$. Where D = diameter.

4. $R = \frac{\rho V}{\pi^2 r^4}$. Where V is the volume.

5. $R_t = R_0(1 + \alpha t)$. Where α = temperature co-efficient of R. Unit of α is per ^{0}C.

6. *To be remembered :*

(i) 1 amp = 10^{-1} e.m.u. of I = 3×10^9 e.s.u. of current.

(ii) 1 coul = 10^{-1} e.m.u. of charge = 3×10^9 e.s.u. of charge.

(iii) 1 volt = 10^8 e.m.u. of potential difference or 10^8 ab bolt

$= \frac{1}{300}$ e.s.u. of potential difference or $\frac{1}{300}$ stat volt.

(iv) 1 ohm = 10^9 e.m.u. of R = $\frac{1}{300}$ stat volt.

7. *Series grouping* : $R_s = r_1 + r_2 + r_3 + \ldots\ldots + r_n$.

The equivalent conductance is given by

$$\frac{1}{G} = \frac{1}{g_1} + \frac{1}{g_2} + \frac{1}{g_3} + \ldots\ldots \frac{1}{g_n}$$

$$V = V_1 + V_2 + V_3 + \ldots\ldots\ldots V_n.$$

In a series potential difference across any resistance (r) is

$$v' = \frac{r'}{r_{eq}} V$$

where r_{eq} = equivalent resistance of series

V = voltage across series, for Ex. – from figure

$$V_1 = \frac{R_1}{R_1 + R_2 + R_3} . V .$$

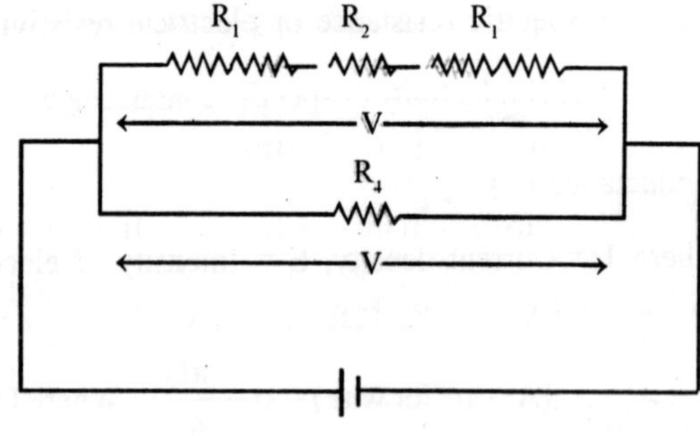

8. *Parallel grouping* :

$$\frac{1}{R_p} = \frac{1}{r_1} + \frac{1}{r_2} + \frac{1}{r_3} + \ldots\ldots + \frac{1}{r_n} .$$

or $G = G_1 + G_2 + G_3 + ... + G_n.$

$I = I_1 + I_2 + I_3 + ... I_n.$

In parallel, current through any Branch

$$i_1 = i \times \frac{R_2}{R_1 + R_2}.$$

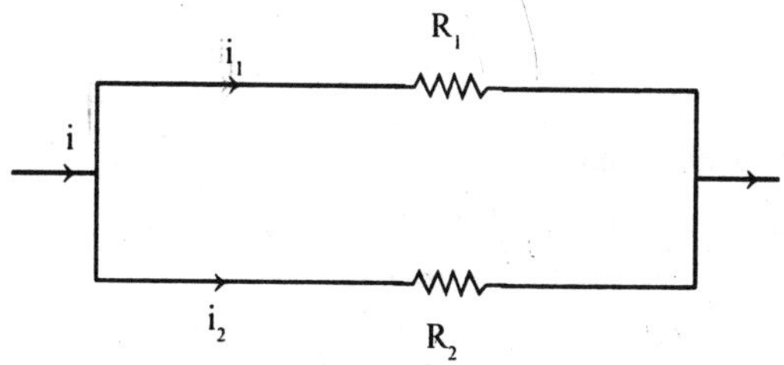

9. *For two resistances in parallel :* $R_p = \frac{r_1 r_2}{r_1 + r_2} = \frac{\text{product}}{\text{sum}}.$

10. *For n identical resistances :* $R_s = nr$ (series).

$$R_p, = \frac{r}{n} \text{ (parallel) and } \frac{R_s}{R_p} = n^2$$

11. Approximate percentage change in R = 2x small percentage changes in length by stretching.

E.M.F., P.D. and Grouping of Cells

1. $\Delta V = \frac{\Delta W}{Q}$ or $\Delta W = Q\Delta V$. S.I. or practical unit of P.D. = Joule/ coul = volt e.s.u. of P.D. = erg/stat coul = stat volt.

2. $1 eV = 1.6 \times 10^{-19}$ Joule. 1 MeV = 1.6×10^{-13} Joule (Million eV).

   1 BeV = 1.6×10^{-10} Joule (Billion eV).

3. $I = \frac{E}{R + r}$ (closed circuit) of E = V + Ir where E = e.m.f.

   V = P.D., r = internal resistance, R = external resistance and Ir = potential drop.

   If, R = 0 (short circuit), I will be maximum, $I_{max} = \frac{E}{r}$.

   If, R = $\propto$ (open circuit), I will be minimum, $I_{min} = 0$.

4. *Series grouping of cells* : $I = \frac{nE}{nr + R}$

$I_{max} = \frac{E}{r}$, where r >> R.

5. *Parallel grouping of cells* : $I = \frac{nE}{r + nR}$; $I_{max} = \frac{nE}{r}$, when r >> R.

6. *Mixed grouping of cells* : $I = \frac{mnE}{nr + mR}$; $I_{max} = \frac{nE}{2R} = \frac{mE}{2r}$, when nr = mR.

Where n = number of cells in one row, m = number of rows and m × n = total number of cells.

7. *Wrong series connection* : $I = \frac{(n - 2m)E}{R + nr}$;

$$I_{max} = \frac{(n - 2m)E}{nr},$$

when R << r. Where n = total number of cells and m = number of cells wrongly connected.

Wheatstone Bridge & Kirchoff's Law

1. When galvanometer shows no deflection, PR = QS (i.e. products of alternate arms resistances are equal) or $\frac{P}{Q} = \frac{S}{R}$.
2. **Krichoff's Laws**

Ist law : $\Sigma I = 0$ 2nd law : $\Sigma I\, R = \Sigma E$.

Where E = total e.m.f. of the mesh or circuit.

Heating Effect of Current and Thermo-Electricity

1. Q = It. Where Q = charge, I = current, t = time.
2. **P.D.** across ends of conductor, V = IR. V = Voltage, I = Current, R = Resistance.
3. *Work done (W)* = Q.V. = VIt = I^2 Rt = $\frac{V^2}{R} \times t = P \times t$.
4. *Power consumption (P)* = $\frac{W}{t} = I\,V = I^2 R = \frac{V^2}{R}$.
5. *Heat produced (H)* = $\frac{W}{J} = \frac{VIt}{J} = \frac{I^2Rt}{J} = \frac{V^2}{R} \times \frac{t}{J} = \frac{P}{J} \times t$.

6. *Energy* = P × t.
7. *Practical unit* of energy supply = killo-watt-how (kWh). 1 I.B.O.T. unit = 1 kWh = 36×10^5 Joules. 1 H.P. = 746 wolt.
8. *Number of units consumed* $= \frac{\text{watt} \times \text{hr}}{1000}$ = kWh.

   S.I. unit : I → amp, V → volt, R → ohm, t → sec, W → Joule, P → watt, H → Joule.

   M.K.S. Unit : 1 cal = 4.2 Joule, H → cal, J = 4.2 Joule/cal.
9. Heat produced H α R if I and t are constant, i.e. $\frac{H_1}{H_2} = \frac{R_1}{R_2}$.
10. Heat produced $H \propto \frac{1}{R}$ if V and t constant, i.e. $\frac{H_1}{H_2} = \frac{R_2}{R_1}$.
11. $H \propto I^2$ if R and t are constant.
12. H α t if R and I are constant.
13. *Hot wire instrument* (for measuring both A.C. & D.C.) :

    $\theta \propto H$ or $\theta \propto i^2$, i.e. $\frac{\theta_1}{\theta_2} = \frac{i_1^2}{i_2^2}$.
14. $t_n = \frac{t_i + t_c}{2}$. Where t_n = neutral temperature t_i = inversion temperature & t_c = temperature of cold junction.

CHEMICAL EFFECT OF CURRENT

1. **Faraday's first law :**

   By flowing a charge Q

   through a electrolytic if mass m of substance liberated or deposited at the electrode

   $m \propto Q \rightarrow m = zQ = Z$ if z = Electrochemical equivalent of substance

   Second law : current flowing through different electrolytes.

   Then

   $m_1 : m_2 : m_3 \,..... = E_1 : E_2 : E_3$

   $\therefore \quad m \propto E$

   E = chemical equivalent of substances.
2. Charge on one ion = ne. Where n = valency of ion, e = electronic charge = -1.6×10^{-19} coulomb.

3. Number of ions liberated at electrode = $\frac{Q}{ne}$. Where Q = charge flowing in electrolyte.

4. Mass of one atom = $\frac{A}{N}$. Where A = atomic weight N = Avogadro's number.

5. Mass of element (w) liberated at electrode = number of ions liberated × mass of one atom.

   i.e. $w = \frac{Q}{ne} \times \frac{A}{N} = \frac{1}{Ne} \times \frac{A}{n} \times Q = \frac{E}{F} \times Q$. Where E = eq. wt.

6. $\frac{E}{F} = Z$ (electrochemical equivalent). S.I. unit of Z is kg/coul.

Shunt, Ammeter, Voltmeter and Branching of Current

1. To increase the range of an ammeter or to convert a galvanometer into an ammeter, a low resistance (shunt) is connected in parallel.

   $S = \frac{G}{n-1}$ (in parallel). $n = \frac{I}{I_g}$.

   Where S = shunt resistance,

   G = resistance of galvanometer

   $\frac{I}{I_g} = 1 + \frac{G}{S}$

   Where I_g = full scale defection current or current which the ammeter can measure.

   I = Current to be measured.

2. To increase the range of a *voltmeter* or to convert an *galvanometer* into a *voltmeter,* a high resistance (R) is connected in series :

   $R = G(n-1) = \frac{V}{Ig} - G$, where G = resistance of galvanometer.

   $n = \frac{V}{V_g}$ or $= \frac{\text{new range}}{\text{old range}}$. Where V_g or V_1 = voltage which the voltmeter can measure and V or V_2 = voltage to be measured.

   Where i_g = current flows through galvanometer, V = voltage to be measured.

3. Branching of current in two parallel resistances :

$$I_1 = \frac{Ir_2}{r_1 + r_2} \text{ and } I_2 = \frac{Ir_2}{r_1 + r_2}.$$

4. After shunting the decrease in resistance $= \frac{G^2}{G+S}$.

ELECTRIC CHARGE AND CURRENT

1. $I = \frac{\Delta Q}{\Delta t}$.
2. Electric charge is a fundamental property.
3. Q = ± ne = It (I = current).
4. Units S.I.–coloumb

   1 coulomb = 6.25×10^{18} electron.
5. Strength of storage cell (A.H.) = charging current (amp) × time (hr). [1 amp × hr = 3600 coulomb].
6. Current density $\vec{j} = \frac{I}{A}$ Amp/m^2.
7. (A = Area of cross section perpendicular to the direction of (I)
8. v_d (drift velocity of electron) =

$$\frac{I}{neA} = \frac{J}{ne} = \frac{\sigma E}{ne} = \frac{E}{\rho\, ne} = \frac{V}{\rho\, l ne}$$

   E = Intensity of electric field

   V = P.D. across conductor.
9. If length of a wire increased by x% then its resistance increases by 2x%.
10. If length of wire becomes n times, It's resistance becomes n^2 times.

Electric Current

(i) Electric current arises due to continuous flow of charged particles or irons.

(ii) Any medium, which have practically free electric charges capable of moving from one place to other, is a conductor of electricity i.e., it can carry an electric current. e.g.

(a) A solid metal consists of-vely charged free electrons and +vely charged ions. Free electrons can migrate from one place to other but +vely charged ions are fixed in their mean positions and can only vibrate about their mean positions This is a solid state conductor, in which motion of electrons constitutes an electric current.

(b) An electrolyte consists of equally and oppositely charged ions (as a result of its dissociation), both of which are capable of moving in the medium. This is a solid state conductor, in which motion of electric current is constituted due to motion of both +ve and- ve ions.

(iii) During the flow of current in a medium, the electric charges always flow from higher potential energy state to lower potential energy state. Therefore, a +ve charge always moves from higher electric potential to lower electric potential while a-ve charge from a lower electric potential to higher electric potential.

(iv) Electric charge flowing per second in an electric circuit is a measure of electric current in that circuit. If a charge q flows through an electric circuit in time t seconds, then electric current in that circuit is

$$i = q/t$$

If a small charge dq flows in time dt, then electric current

$$i = dq/dt$$

(v) The direction of current is *conventionally* taken as the direction of flow of +ve charge. If in a medium, current is due to flow of -ve charges, then direction of current will be opposite to direction of motion of-ve charges.

(ii) The unit of electric current in SI system is Ampere.

$$1 \text{ Ampere} = \frac{1 \text{ coulomb}}{1 \text{ second}}$$

(vii) Because in metals, current is due to flow of electrons and charge on one electron is equal to 1.6×10^{-19} coulomb, hence for one coulomb of charge $[(1/1.6 \times 10^{-19}) = (6.25 \times 10^{18})]$ electrons must flow i.e.

1 Ampere = 6.25×10^{18} electrons/sec.

This means that in one second 6.25×10^{18} electrons enter at one end of wire and same number of other electrons leave at the other end.

Current Density

(i) The current density at any point inside a conductor is defined as a vector quantity, having magnitude equal to current through a very small unit area around that point, the area being normal to direction of flow of current and direction along the direction of current at that point.

(ii) If A_n represent a small area at the point P, normal to direction of current 1, then

Current density $J = \frac{I}{A_n}$ ...(1)

If the plane of small area is not normal to the current but makes an angle θ with the area normal to current, then,

$J = \frac{I}{A_n} = \frac{I}{A \cos \theta}$ ...(2)

(iii) Unit of current density in SI system is ampere/meter^2.

(iv) Because current density and area are vector quantities, hence equation (2) can also be written as

$I = JA \cos \theta = \vec{J}.\vec{A}$ ...(3)

ELECTROLYSIS (CHEMICAL EFFECT OF CURRENT)

(i) It has been found experimentally that when certain liquids, bases, salts etc. are dissolved in water, they get dissociated and become conductors of electricity e.g. NaOH dissolved in water, $CuSO_4$ dissolved in water, $AgNO_3$ dissolved in water etc. Charge carriers responsible for electric conduction are positve and negative ions.

(ii) The phenomenon of decomposition of substance into + ve and – ve ions, due to passage of electric current in its solution, is known as *Electrolysis* or *Chemical Effect of Electric Current.*

(iii) The liquids in which electrolysis takes place, are called as *Electrolytes.* The plate immeresed in electrolyte for the passage of current are called as *Electrodes.* The plate at which current enters the electrolyte is called the *Anode* and that at which current leaves the electrolyte is called the *Cathode.* The contaainer in which the process of electrolysis is carried out is called as *Voltameter..*

(iv) Voltameter in which electrolyte is copper sulphate solution is called the *Copper voltameter,* and that having silver nitrate solution is called the *Silver voltameter.*

(v) Ions moving in a Voltameter towards Cathode and Anode are called as Cations and Anions respectively.

(vi) Cations move in the direction of electric field while anions (having –ve charge) move in a direction opposite to that of electric field.

(vii) Because direction of current is taken as direction of motion of +ve charge and +ve and –ve ions move in opposite directions, hence current due to both types of ions is in same direction i.e. along the direciton of electric field.

Faraday's Laws of Electrolysis

Faraday carried out experiments on different electrolytes and proposed following two laws of electrolysis on the basis of his observations:

First Law : The mass of an element liberated at an electrode in the process of electrolysis, is directly proportional to the quantity of electricity of total electric charge passed through the electrolyte. If m kg is the mass of a substance liberated due to flow of a current of i ampere for t seconds, then

$m \propto q$ (where q represents the quantity of charge flown in t sec)

Because $q = it$, hence $m \propto it$

or $\mathbf{m = Zit}$

Important Points

(a) Z is constant for the given element, called as *Electro chemical equivalent* or *E.C.E.*

(b) If i = 1 sec, Then Z = m i.e. E.C.E. of an element is equal to the mass of the element liberated when one ampere of current is passed through the electrolyte for one sec.

(c) $Z_{Ag} = 1.118 \times 10^{-6}$ Kg/coulomb;

$Z_{Cu} = 0.329 \times 10^{-6}$ Kg/coulomb

(d) Unit of Z in SI system = Kg/coulomb

(e) $[Z] = [m]/[q] = [MA^{-1}T^{-1}]$

(f) **International ampere :** One international ampere is that value of electric current which when passed through a solution of silver nitrate for one second, liberates 1.118×10^{-6} Kg of silver.

Second law : If same quantity of electricity is passed through different electrolytes simultaneously (by joining different voltameters in series), then masses of different substances liberated are found to be in the direct ratio of their chemical equivalents.

Note : If the atomic weight of an element be A and its valency be V, then its chemical equivalent is given by

$W = A/V$ e.g. $W_{Ag} = 107.9/1 = 107.9$,

$W_{Cu} = 63.5/2 = 31.75$

When chemical equivalent is expressed in Kg, then it is called Kg-Equivalent,

e.g. 1 kg - equivalent of Ag = 107.9 Kg.

1 kg – equivalent of Cu = 31.75 Kg.

If masses of elements liberated are m_1 and m_2 due to passage of same quantity of charge through two different electrolytes simultaneously and W_1 & W_2 be their chemical equivalents respectively, then according to Faraday's 2nd law

$$\frac{m_1}{m_2} = \frac{W_1}{W_2}$$

e.g. If same charge is allowed to pass through water, copper and silver voltameters simultaneously, then hydrogen, copper and silver liberated at the cathodes will be in the ratio 1 : 31.75 : 108 respectively i.e. in the ratio of their chemical equivalents.

Faraday Constant

According to Faraday's second law,

we know that $$\frac{m_1}{m_2} = \frac{W_1}{W_2} \quad ...(1)$$

If Z_1 and Z_2 are the electrochemical equivalents of two elements, then according to Faraday's 1st law, we have $m_1 = Z_1$ it and $m_2 = Z_2$ it

$$\frac{m_1}{m_2} = \frac{Z_1 \text{ it}}{Z_2 \text{ it}} = \frac{Z_1}{Z_2} \quad ...(2)$$

Comparing equation (1) & (2)

$$\frac{Z_1}{Z_2} = \frac{W_1}{W_2} \text{ or } \frac{W_1}{Z_1} = \frac{W_2}{W_2}$$

$$\text{or } \frac{W}{Z} = \text{A Constant} \qquad ...(3)$$

Because the ratio W/Z is same for all the elements, hence it is called as Faraday's constant F.

For Example :

(i) $$F_{Ag} = \frac{W_{Ag}}{Z_{Ag}} = \frac{107.9 \text{ Kg}}{1.118 \times 10^{-6} \text{ Kg/Coulomb}}$$

$$= 9.65 \times 10^7 \text{ Coulomb/Kg-Equivalent}$$

(ii) $$F_{Cu} = \frac{W_{Cu}}{Z_{Cu}} = \frac{31.75 \text{ Kg}}{0.329 \times 10^{-6} \text{ Kg/Coulomb}}$$

$$= 9.65 \times 10^7 \text{ Coulomb/Kg-equivalent}$$

Thus *Faraday's constant is equal to the quantity of charge required to liberate one kg - equivalent of any substance by electrolysis.* Its value is 9.65×10^7 Coulomb/Kg-equivalent.

Note : Charge required to liberate 1 gm - equivalent of any substance = 96,500 coulomb = 1 Faraday.

Relation between Faraday's constant, Avogadro's number and fundamental charge

As charge required to liberate 1 kg – equivalent of any substance by electrolysis = F

Hence, charge required to liberate 1 kg-atom of any substance of valency V = FV.

According to Avogadro's Hypothesis, 1 kg. atom of any substance consists of N (= 6.02×10^{26}) atooms. The number N is known as Avagadro's number.

Hence, charge required to deposit Natoms of any substance = FV i.e. charge required to deposit one atom of a substance = FV/N.

In the process of electrolysis, atoms liberated at the electrode reach there in the form of ions and after giving their charge to the electrode, are liberated in the form of atoms.

i.e. If, for example, a Cu^{++} ion reaches the cathode, then it is liberated after giving 2e charge to the electrode i.e. charge required to liberate one atom of valency V = Ve

Therefore, (FV/N) = Ve or $F = Ne$.

Conduction in Metals

(i) Metals are very good conductors of electricity among solids. In any atom of a substance, the electrons in the orbits nearer to nucleus are tightly bound to the nucleus but the electrons in the outer orbits are very loosely bound and can be easily detatched from the atoms. These electrons move freely in the vacant spaces between the atoms and are known as conduction electrons of Free electrons. It is these electrons which act as charge carries for conduction of electricity in metals. Because metals are found to have a very high number of these electrons, hence they are good conductors of electricity. *Silver is best conductor of electricity.*

Note : Electric conduction takes place in certain liquids and gases also but there is an important difference between electric conduction in metals and that found in liquids and gases. In metals, it is ionly the –ve charges which take part in conduction while in liquids and gases, both –ve and +ve charges are responsible for conduction of electricity.

(ii) **Drift Velocity**

(a) As the free electrons in metals behave in the same way as do the molecules of a gas, hence they are said to form *Electron gas.* Just like the molecules of a gas, these electrons are also in a state of random motion. They constantly collide with each other and go on changing their direction of motion i.e. they have no net motion in any particular direction i.e. net rate of flow of charge through any plane in the metal is zero i.e. there is no current in the metal.

(b) When we connect the ends of a metallic wire with the terminals of a battery, then an electric field is produced at every point of the wire. Due to this electric field, every free electron now experience a force and gets accelerated in a particular direction i.e. opposite to that of electric field. This leads to establishment of current in the wire.

(c) But the electrons are not accelerated indefinitely but for a very short time. After a momentary acceleration, the electron always collides with a positive ion and loses its energy obtained from battery in the form of heat energy. It is then again accelerated between two positive ions and again collides with next positive ion and so on. *Thus, applied potential difference does not give an accelerated motion to the electrons but only gives them a very small constant velocity which gets superposed on the random motion of electrons.* This small constant velocity is known as *Drift velocity.*

ELECTRIC RESISTANCE

(i) Whenever a potential difference is established between the ends of a conductor, an electric current starts flowing in it. The ratio of potential difference applied across the conductor and the current flowing in it is called as Electric Resistance R of the conductor. If V is the potential difference applied across the conductor and i is the current flowing in it, then electric resistance R is : $R = V/i$

(ii) Unit of resistance in SI system is ohm;
1 ohm = [1 Volt/ 1 Ampere]

(iii) 1 megha ohm = 1 M Ω = 10^6 ohms;
1 micro ohm = 1 μ Ω = 10^{-6} ohms
1 milli ohm = 1 m Ω = 10^{-3} ohms

(iv) Reciprocal of Electric Resistance is called electric conductance and its unit is ohm^{-1} or Mho.

(v) $$[R] = \frac{[V]}{[I]} = \frac{[W/Q]}{[I\}} = \frac{[W]}{[I]\,[Q]} = \frac{[ML^2T^{-2}]}{[A]\,[AT]}$$
$$= [ML^2T^{-3}A^{-2}]$$

Ohm's Law

(i) If the physical state of a conductor remains unchanged (i.e. temperature, material and dimensions), then the ration of the potential difference applied across its ends to the current flow in it, remains constant. i.e. If V is the potential difference applied across the ends of the conductor and i is the current flowing through it, then according to ohm's law

$$\frac{V}{i} = \text{A Constant}$$

(ii) Because, the ratio V/i is called as the electrical resistance R of the conductor, hence

$$\frac{V}{i} = \text{A Constant} = R$$

Hence ohm's law can now be expressed as – If the physical state of a conductor remains unchanged, then Resistance of the conductor remains constant, whatever may be the potential difference applied across it. In other words, a graph plotted between applied potential difference V and the current i flowing through the conductor will always be a straight line.

Note : Ohm's law is valid for metallic conductors only.

(iii) **Proof of Ohm's Law**

Let us consider a conductor of length/and cross sectional area A. Suppose a current i flows through the conductor when a potential difference V is applied across its ends. If n is the number of free electrons per unit volume and v_d is the drift velocity of electrons, then we know that

$$i = neAv_d \qquad ...(1)$$

Because V is potential difference applied across the conductor of length l, hence electric field produced at every point of the conductor is given by $E = V/l$

Hence electric force acting on every free electron, is

$$F = eE = \frac{eV}{l}$$

If m is the mass of electron, then acceleration experienced by the electron due to this force

$$a = \frac{F}{m} = \frac{eV}{ml} \qquad ...(2)$$

The electrons are not accelerated continuously but for a very short interval of time, called as Relaxtion time τ (= time taken between two successive collisions). Hence, velocity acquired by electron during relaxation time τ is given by

$v = 0 + a\tau = a\tau = (e\ v\ \tau/ml)$

After every collision with +ve ion, the velocity of electron becomes zero and again increases to v till it again collides with next +ve ion and so on. Hence average or drift velocity of electrons :

$$v_d = \frac{0+v}{2} = \frac{v}{2} = \frac{eV\tau}{2ml} \qquad ...(3)$$

Substituting eq. (3) in eq. (1), we get

$$i = neA\frac{eV\tau}{2ml} = \left[\frac{ne^2\tau}{2m}\right]\frac{A}{l}V$$

or

$$\frac{V}{i} = \left[\frac{2m}{ne^2\tau}\right]\frac{l}{A}$$

For a given metallic wire at a given temperature, all the quantities on R.H.S. of above equation will be constant. This mean that (V/i) = R = constant

This thus proves *ohm's law.*

Resistivity or Specific Resistance

(i) We know that Resistance of a conductor of length l and C.S. area A is given by

$$R = \left[\frac{2m}{ne^2\tau}\right]\frac{l}{A}$$

Here quantities within bracket depend upon temperature and material of the conductor but those outside the bracket depend only on the dimensions of the conductor. Hence, if temperature and material of the conductor are kept constant, then quantity within the bracket will be a constant. This constant quantity is called as specific resistance and is represented by ρ.

(ii) Specific Resistance depends only on temperature and material of the conductor but not on its dimensions. As ρ depends only on the material of a conductor at a given temperature, hence it is a *characteristic constant.*

(iii) Hence $R = (\rho l/A)$ or $\rho = (RA/l)$;

If A = 1 and l = 1, then ρ = R

i.e. the specific resistance of the material of a conductor is equal to the Resistance of a conductor of that material, having unit length and unit cross sectional area, at a given temperature.

(iv) Alternatively, the specific resistance of a material is also defined as the resistance offered by an Unit Cube of that material between its two opposite faces, when current enters ⊥ to one face of cube and leaves ⊥ to opposite face.

(v) Unit of ρ in SI system = ohms x meter

(vi) $[\rho] = \frac{[R][A]}{[l]} = \frac{[V][A]}{[I][l]} = \frac{[W][A]}{[q][I][l]}$

$= \frac{[ML^2T^{-2}][L^2]}{[AT][A][L]} = [ML^3T^{-3}A^{-2}]$

(vii) **Relation between ρ, E and J**

$\rho = R\frac{A}{l} = \frac{V}{I}\times\frac{A}{l} = \frac{V/l}{I/A} = \frac{E}{J}$

(viii) **Specific conductance**

(a) Reciprocal of specific resistance is called as specific conductance and is represented by σ

(b) $\sigma = \frac{1}{\rho} = \frac{J}{E}$

or J = σ E

(c) Unit of s is ohm^{-1} $meter^{-1}$

(xi) The property of specific resistance of the materials can be used to select the proper material for making resistance wires needed for different purpose.

(a) *For making connection wires, we must use materials having low specific resistance* so that resistance of connection wires may be minimum possible and they may not effect the current in a circuit in which they are used.

As ρ for pure metals like silver, copper, gold, aluminium etc. is quite low, hence these materials are used for making connection wires.

(b) *For making resistance wires used for making resistance boxes, we must use materials having high specific resistance* so that a given resistance may be obtained by using a smaller length of the wire.

As specific resistance of the alloys like Constantan, Nichrome, Mangnin etc. is quite high, hence these alloys are used for making resistance wires used in resistance boxes.

Note : These alloys have one more special property that their *temperature coefficient of resistance (α) is quite.* Because of this property, the resistance of a Resistance box is negligibly effected by change of temperatue.

Some Important Points Concerning Resistance

(i) We know that $R = \left[\frac{2m}{ne^2\tau}\right]\frac{l}{A}$

For a given material of the conductor at a given temperature – R ∞ l and R ∞ (1/A)

Hence Resistance of a long and thin wire will be greater than that of a short and thick wire.

(ii) If a wire of a given resistance R is stretched to make its length doubled then its resistance becomes four times of its previous value.

Because volume will remain constant in the process of stretching, hence $Al = A'2l$ i.e. $A' = A/2$ i.e. New resistance of the stretched wire

$$R' = \rho\frac{l'}{A'} = \rho\frac{2l}{A/2} = 4\rho\frac{l}{A} = 4R$$

(iii) If a wire of given resistance R is melted and a fresh wire of double radius is prepared, then resistance of new wire is reduced to 1/16 of the previous value.

Again, because volume will remain constant, hence $Al = 4\,Al'$ i.e. $l' = l/4$

i.e. resistance of fresh wire

$$R' = \rho\frac{l'}{A'} = \rho\left[\frac{l/4}{4A}\right] = \frac{1}{16}\left[\frac{\rho l}{A}\right] = \frac{R}{16}$$

EFFECT OF TEMPERATURE ON RESISTANCE

(a) Resistance of Pure Metals

(i) We know that $R = \left[\frac{2m}{ne^2\tau}\right]\frac{l}{A}$

For a given conductor, l, A and n are constant, hence

$R \propto (1/\tau)$

If γ represents the mean free path (Average distance covered between two successive collisions) of the electron and v_{rms}, the root - mean - square speed, then

$$\tau = \frac{\lambda}{v_{rms}}, \text{ Hence } R \propto \frac{v_{rms}}{\lambda}$$

(a) λ decreases with rise in temperature because the amplitude of vibrations of the +ve ions of the metal increases and they create more hindrance in the movement of electrons and ,

(b) v_{rms} increase because $v_{rms} \propto \sqrt{T}$. Therefore, *Resistance of the metallic wire increases with rise in temperature.* As $\rho \propto R$ and $\sigma \propto (1/\rho)$, hence resistivity increase and conductivity decrease with rise in temperature of the metallic wires.

(ii) If R_0 and R_1 represent the resistances of metallic wire at 0°C and t °C respectively then R_1 is given by the following formula:

$$R_1 = R_0 (1 + \alpha t) \qquad ...(1)$$

where α is called as the *temperature coefficient of resistance* of the material of the wire.

From equation (1) we can write

$$\alpha = \frac{R_t - R_0}{R_0 \times t} \text{ per } ^\circ C$$

If R_0 = 1ohms, t = 1 ° C, Then $\alpha = R_t - R_0$ i.e. the temperture coefficient of resistance of the material of a wire is equal to increase in its resistance on raising its temperature by 1°C if its resistance at 0°C is assumed to be 1 ohm.

(iii) For most of the metals, α is nearly equal to (1/273) per °C, hence according to equation (1)

$$R_t = R_0\left[1 + \frac{t}{273}\right] = R_0\left[\frac{273+t}{273}\right] = R_0\left[\frac{T}{273}\right]$$

where T is the absolute temperature. Hence $R_t \propto T$

i.e. *Resistance of pure metallic wires or the resistivity of metals is directly proportional to absolute temperature.*

(b) Resistance of Alloys

(i) Resistance or resistivity of alloys also increases with increase in temperature but the increment in Resistance is found to be much smaller as compared to pure metals.

(ii) *α for alloys is +ve and very small*

(iii) There are some special alloys such as constant, mangnin etc. whose temterature coefficient of resistance is negligible i.e. whose resistivity is negligibly effected by rise in temperature. As mentioned earlier also, it is because of their high resistivity and low temperature coefficient of resistance that these materials are used for making resistance wires used in standard resistance boxes.

(c) Resistance of Semiconductors

(i) There are certain substances whose conductivity lies in between that of insulators and conductors, higher than that of insulators but lower than that of conductors. These are called as semiconductors, e.g. silicon, germanium, carbon etc.

(ii) The resistivity of semiconductors decreases with increases in temperature i.e. *α for semicondutors is –ve and high.*

(iii) Though at ordinary temperature the value of n (no. of free electrons per unit volume) for these materials is very small as compared to metals, but increases very rapidly with rise in temperature (this happens due to breaking of covalent bonds). Therefore, the resistance

$$R = \frac{2ml}{ne^2 \tau A}$$ goes on decreasing with increase in temperature.

Note : It is worth mentioning that here also τ decreases with rise in temperature but increase in the value of n is much higher as compared to decrease in τ i.e. net effect is that resistance decreases with rise in temperature.

(d) Resistance of Electrolytes

Resistance or resistivity of electrolytes also decreases with rise in temperature. This all happens due to the fact that viscosity of electrolytes decreases with rise in temperature, so that ions now get more freedom of motion i.e. Resistance decreases with rise in temperature.

(e) Resistance of Superconductors

There are also some such substances whose resistance shows a special behaviour with change in temperature, in low temperature region. As the temperature of these substances is decreased, their resistance first decreases in ordinary way, just like pure metals, but after a certain fixed very low temperature the resistance decreases extremely rapidly and tends to zero e.g. resistance of mercury decreases very rapidly and tends to zero at 4°K. This phenomenon is called as superconductivity.

GENERATION OF HEAT BY ELECTRIC ENERGY

(i) When the ends of a metallic wire are connected with the terminals of a battery, free electrons present in the conductor starts moving with a drift velocity in a direction opposite to that of applied electric field and a current is established in the conductor. These electrons are not accelerated indefinitely but constantly collide with +ve ions present in the conductor and transfer their energy obtained from the battery, to these +ve ions. Thus electrons go on taking energy from the battery and losing it continuously in undergoing collisions with +ve ions. As a result of this, root mean square velocity of ions is increased i.e. temperature of wire is increased i.e. electrical energy is dissipated in the form of heat energy.

(ii) Let a current of i ampere is allowed to flow in a metallic wire for t seconds when a potential differences V is established between two ends of wire. If the charge q flows through the wire in t seconds, then q it.

Hence work done by the battery in taking q coulomb of charge from one end of wire to other, is given by

$$W = Vq = Vit \text{ Joules} \qquad ...(1)$$

If R represents the resistance of metallic wire, then $V = iR$ or $i = (V/R)$

$$\text{Hence } W = i^2 Rt = (V^2t/R) \text{ Joules} \qquad ...(2)$$

$$\text{Thus, } W = Vit = i^2 Rt = (V^2t/R) \qquad ...(3)$$

Because 1 calorie of heat is equivalent to 4.2 Joules of work, hence heat produced or energy dissipated in the metallic wire in t sec is given by

$$H = \frac{W}{4.2} = \frac{Vit}{4.2} = \frac{i^2 Rt}{4.2} = \frac{V^2 t}{4.2 R}$$

ELECTRIC POWER

(i) The rate at which electric energy is consumed in an electric circuit, is called as its Electric power and is represented by P.

(ii) Hence from (3) we can write :

$$P = Vi = i^2 R = (V^2/R) \quad ...(4)$$

(iii) The unit of power is *Watt.* If in an electric circuit, energy is dissipated at the rate of 1 Joule/sec, then the power of the circuit is 1 Watt i.e. 1 Watt = (1 Joule/1 sec)

We can also write from eq. (4) –

1 Watt = 1 Volt × 1 Ampere

(iv) 1 KW = 1 Kilo Watt = 10^3 Watt,

1 MW = 1 Megha Watt = 10^6 Watt

1 mW = 1 milli Watt = 10^{-3},

Also, 1 Horse power = 746 Watt.

(v) It is also clear from eq. (4), *that at constant temperature, rate of production of heat in a current-carrying wire is either directly proportional to the square of the current or to the square of the potential difference.* This is known as Joule's law.

Note : It is to be noted here that if ohm's law remains valid, Joule's law also remains valid.

Some important conclusions from the concept of Electric Power

(i) If 100W-220 V is written on some electric instrument, then it simply means that on using the instrument on 220 volt supply, 100 Joules of energy is consumed by it one second.

(ii) *Other informations from voltage and wattage* : Let us take an example of a bulb whose wattage is P watts and which is manufactured for working on a supply of V volts (which is constant and equal to 220 volts). From this data we can find out the following-

(a) *Resistance of the filament of bulb*

$$R = V^2/P$$

As V is constant, hence *higher is the Wattage of a Bulb, lesser is the resistance of its filament or thicker is the filament* ($R \propto 1/A$).

(b) *Maximum current that can be allowed to pass through a bulb-*

$i_{max} = P/V$

i.e. higher is the wattage of bulb, higher is current that can be allowed to pass through a bulb.

Note : Value of i_{max} is fixed for a given bulb. If a bulb is being used in a circuit and current in that circuit is higher than i_{max} for that bulb, then bulb get fused.

(iii) If bulbs of different wattage are joined in parallel, then highest wattage bulb glows with maximum brightness.

Reason : As bulbs are joined in parallel, hence voltage across each bulb will be same. Now, because resistance of highest Wattage bulb is minimum, hence *heat produced (= $V^2 t/4.2R$) (or brightness) will be maximum in highest wattage bulb.*

This is the reason that connections in houses are made in parallel.

Note : There is one more advantage of making connections in parallel. If any branch of the circuit goes out of order, then rest of the circuit will go on working. Moreover, there will be an independent on-off switch for every appliance being used in the circuit of a house.

(iv) If bulbs of different wattage are joined in series, then lowest wattage bulb glows with maximum brightness.

Reason : As bulbs are joined in series, hence current through each bulb will be same. Now because resistance of lowest wattage bulb is maximum, hence *heat produced (= $i^2 Rt/4.2$) will be maximum in lowest Wattage bulb.*

Kilowatt-hour

(i) It is a commerical unit used for expressing consumed electric energy.

(ii) 1 KWH or 1 unit is the quantity of electric energy consumed in one hour in an electric circuit having power of 1 Kilo Watt.

(iii) 1 KWH = 1000 watt × 1 hour = [(1000 Joules/second) × (3600 seconds)] = *36 × 10^5 Joules*

(iv) *Number of units consumed in an electric circuit*

If a current of i ampere flows in a circuit for t hours under a potential difference of V volts, then electric energy dissipated in the circuit = Electric power × time

= Vi (Watt) × t (hours)

= [(Vi/1000) (Kilo Watt)] × t (hours)

= (Vit/1000) Kilo Watt hour

∴ number of units

$$= \frac{\text{Volt} \times \text{Ampere} \times \text{hours}}{1000} = \frac{\text{Watt} \times \text{hours}}{1000}$$

$$= \frac{(\text{Ampere})^2 \times \text{ohm} \times \text{hours}}{1000}$$

E.M.F. of a Cell

We know that when electrons move through a conductor, their electrical energy is always consumed in undergoing collisions with ions of the conductor. Hence in order to maintain the flow of charge or electric current in an electric circuit continuously, some external agency is required for providing energy to the electrons for flowing through the whole circuit. The cell plays the role of this external agency. The energy liberated in the cell due to chemical reactions taking place in it, maintains the flow of charge in the circuit. In this way, chemical energy of the cell gets converted into electrical energy which in turn is dissipated in the form of heat energy. *Whatever work is done by the cell (or energy is given by the cell) for flow of unit charge throughout the whole circuit (including the cell) is called as E.M.F. of the cell.*

If W is the work done by the cell for flow of charge q throughout the whole circuit, then E.M.F. of a cell E = W/q

Potential Difference

Suppose a source of constant e.m.f. E is connected in a circuit consuming of three resistances R_1 R_2 & R_3 joined end to end. If W is the energy given by the cell for flow of charge q throughout the circuit, then E = W/q.

If W_1, W_2 and W_3 are the energies consumed in different parts of the circuit due to flow of charge q, then $W = W_1 + W_2 + W_3$.

and $E = \frac{W}{q} = \frac{W_1 + W_2 + W_3}{q} = \frac{W_1}{q} + \frac{W_2}{q} + \frac{W_3}{q}$

or $E = V_1 + V_2 + V_3$

where $V_1 = \frac{W_1}{q}$, $V_2 = \frac{W_2}{q}$ and $V_3 = \frac{W_3}{q}$

clearly represent the energies consumed in different parts of external circuit due to flow of an unit charge. These are known as *Potential Differences* across different parts of external circuit. Potential difference is measured by voltmeter.

Note : The potential difference between the two terminals of a surce, when no energy is drawn from it,is called as its Electromotive Force or in other words emf of a source is equal to the potential difference between its terminals when no current is drawn from it or when the source or cell is in open circuit.

Internal Resistance of a Cell

When the terminals of a cell are connected by a wire, an electric current flows in the wire from positive terminal of the cell towards the negative terminal, but in side the electrolyte of the cell it flows from the negative terminal to positive terminal. Just as, we have read, material of the wire opposes the flow of current, the electrolyte of the cell also creates opposition or resistance in the path of current. This resistance, due to the electrolyte, is known as Internal resistance of the cell. Due to this resistance , a part of the energy supplied by the cell is dissipated in the cell itself in the form of heat energy.

Therefore work done by the cell or energy given by the cell for flow of charge q, is

$W = Eq = Eit.$

This energy is used up in two ways.

(a) *In external circuit :* If V is the potential difference across the external resistance R (which is also the potential difference between terminals of cell, when current is being drawn from the cell) then energy consumed in it (or work done outside the cell) is

$W_{ext} = Vit$

(b) *In internal circuit :* If V' is the potential drop in the electrolyte of the cell due to its internal resistance r, then energy consumed inside the cell (or work done inside the cell) is

$W_{int} = V' i t = i^2 r t$

Hence, according to law of conservation of energy

$W = W_{ext} + W_{int}$

$Eit = Vit + i^2 r t$

$E = V + ir$

or $V = E - ir$ ...(1)

Important Points

(i) If a cell is being discharge or current is being drawn from the cell, then potential difference across the terminals of cell is always less than its E.M.F. It happens due to a potential drop across the internal resistance of the cell.

(ii) Higher is the value of current drawn from the cell, lesser will be the potential difference across the terminals of the cell.

(iii) EMF is the characteristic property of the cell which remains constant for a cell but potential difference goes on decreasing as more & more current is drawn from the cell, due to increase in potential drop across the internal resistance.

(iv) If the cell is in open circuit or no current is being drawn from the cell, then from eq. (1) we get V = E i.e. potential difference between the terminals of cell is equal to E.M.F. of the cell.

Note : (a) Generally, a voltmeter is used to measure the E.M.F. of a cell. But when a voltmeter is connected across the terminals of cell, cell no more remains in open circuit because some current is drawn by the voltmeter. As a result of this, reading of voltmeter $<$ E.M.F. of the cell.

(b) However, because of the resistance of voltmeter being quite high, current drawn by voltmeter is very small and therefore reading of voltmeter $\approx$ E.M.F. of the cell.

(c) E.M.F. can be measured accurately with the help of a potentiometer.

(v) If the cell is short circuited, then a maximum current is drawn from the cell (= E/r) and potential difference across its terminals becomes zero.

$$(V = E - ir = E - (E/r) \times r = 0)$$

(vi) If V is the potential difference across the external resistance R due to flow of current i then V = i R and according to eq. (1)

$$iR = E - ir$$

or $$E = i(R + r)$$

or $$i = \frac{E}{R + T} \quad ...(2)$$

If there are more than one cells and external resistances in the circuit, then equation (2) may be written in following generalised form

$$\text{Main current in a circuit} = \frac{\text{Net E.M.F.}}{\text{Total External Resistance} + \text{Total Internal Resistance}} \quad ...(3)$$

(vii) From eq (2) we can write :

$$\frac{V}{R} = i = \frac{E}{R+r}$$

or $$\frac{E}{V} = \frac{R+r}{R} = 1 + \frac{r}{R}$$

or $$r = R\left(\frac{E}{V} - 1\right) \quad ...(4)$$

Thus, r can be measured just by measuring the potential difference across the terminals of cell in two cases.

(a) when key K is open, then reading of voltmeter $\cong$ E.

(b) when key K is closed, then reading of voltmeter $\cong$ V.

KIRCHOFF'S LAWS

Kirchoff gave following two laws for determining the distribution of currents in different branches of a complicated electrical circuit.

First Law : *In a network of conductors, the algebric sum of currents meeting at any junction in the circuit is always equal to zero i.e.*

$$\Sigma i = 0.$$

While applying this law we have to follow a convention for deciding the sign of current. The currents approaching the junction are taken as positive while those leaving the junction are taken as negative. For example, in the figure, currents i_1 i_2 i_6 and i_7 are to be taken as positive while i_3 i_4 and i_5 as negative. Hence according to Kirchoff's 1st law

$$i_1 + i_2 + i_6 + i_7 - i_3 - i_4 - i_5 = 0$$

or $i_1 + i_2 + i_6 + i_7 = i_3 + i_4 + i_5$.

Thus, sum of currents approaching the junction is equal to the sum of currents leaving the junction or total charge approaching the junction per second is equal to that leaving the junction per second i.e. when a steady current flows in the circuit, then there is neither any collection of charge at any point in the circuit nor any charge is removed from there. In other words, *Kirchoff's first law is an alternative form of the law of conservation of charge.*

Second law : *In any closed mesh of a circuit, the algebric sum of the products of the currents and resistances of the different parts of the mesh is always equal to the algebric sum of different emf's acting in that mesh*

i.e. $\Sigma\, i\, R = \Sigma\, E$

Applying Kirchoff's 2nd law to meshes 1 and 2, we get

$$i_1 r_1 - i_2 r_2 = E_1 - E_2 \quad ...(1)$$

and $$i_2 r_2 + (i_1 + i_2)\, R = E_1 \quad ...(2)$$

Solving these equations, we can find currents in different branches of the circuit.

COMBINATIONS OF RESISTANCES

A number of resistances can be connected in a circuit and any complicated combination can be, in general, reduced essentially to two different types, namely series and parallel combinations.

(i) In this combination current flowing through each resistance will be same and will be equal to current supplied by the battery.

(ii) As resistances are different and current flowing through them is same, hence potential differences across them will be different. Applied potential difference will be distributed among three resistances directly in their ratio.

As i is constant, hence $V \propto R$

i.e. $V_1 = iR_1$, $V_2 = iR_2$, $V_3 = iR_3$.

(iii) If the potential difference between the points A and D is V, then

$V = V_1 + V_2 + V_3 = i\,(R_1 + R_2 + R_3)$

(iv) If the combination of resistances between two points is replaced by a single resistance R such that there is no change in the current of the circuit and in the potential difference between those two points, then the single resistance R will be equivalent to combination and $V = i\,R$ i.e.

$iR = i\,(R_1 + R_2 + R_3)$

or $\mathbf{R = R_1 + R_2 + R_3}$.

(v) Thus in series combination of resistances, important conclusions are :

(a) Equivalent Resistance > highest individual resistance.

(b) Current supplied by source = Current in each resistance

$$\text{or} \quad \frac{V}{R_1 + R_2 + R_3} = \frac{V_1}{R_1} = \frac{V_2}{R_2} = \frac{V_3}{R_3}$$

(c) The total potential difference V between points A and B is shared among the three resistances directly in their ratio.

$V_1 : V_2 : V_3 = R_1 : R_2 : R_3$.

Resistances in Parallel

(i) When two or more resistances are combined in such a way that their first ends are connected to one terminal of the battery while other ends are connected to other terminal, then they are said to be connected in parallel. Three resistances R_1 , R_2 and R_3 joined in parallel between two points A and B. Suppose the current flowing from the battery is i. This current gets divided into three parts at the junction A. Let the currents in three resistances R_1 , R_2 and R_3 are i_1 , i_2 i_3 respectively.

(ii) Suppose potential difference between points A and B is V. Because each resistance is connected between same two points A and B, hence potential difference across each resistance will be same and will be equal to applied potential difference V.

(iii) Since potential difference across each resistance is same, hence current approaching the junction A is divided among three resistances reciprocally in their ratio.

As V is constant, hence i ∞ (1/R) i.e.

$$i_1 = \frac{V}{R_1}, i_2 = \frac{V}{R_2} \text{ and } i_3 = \frac{V}{R_3}$$

(iv) Because i is the main current which is divided into three parts i_1, i_2 and i_3 at the junction A, hence

$$i = i_1 + i_2 + i_3 = V\left[\frac{1}{R_1}+\frac{1}{R_2}+\frac{1}{R_3}\right]$$

If the equivalent resistance between the points A and B is R, then i = V/R

$$\text{Thus, } \frac{V}{R} = V\left[\frac{1}{R_1}+\frac{1}{R_2}+\frac{1}{R_3}\right]$$

$$\text{or } \mathbf{\frac{1}{R} = \frac{1}{R_1}+\frac{1}{R_2}+\frac{1}{R_3}}$$

(v) Thus in parallel combination of resistances important conclusions are :

(a) Equivalent Resistance < lowest individual resistance

(b) Applied potential difference = Potential difference across each resistance.

or $iR = i_1R_1 = i_2R_2 = i_3R_3$.

(c) Current approaching the junction A = Current leaving the junction B and current is shared among the three resistances in the inverse ratio of resistances

$$i1 : i2 : i3 = \frac{1}{R_1} : \frac{1}{R_2} : \frac{1}{R_3}$$

WHEATSTONE'S BRIDGE

(i) Wheatstone designed a network of four resistances with the help of which the resistance of a given conductor can be measured. Such a network of resistances is known as Wheatstone's bridge.

(ii) When key K_1 is pressed, a current i flows from the cell. On reaching the junction A, the current i gets divided into two parts

i_1 and i_2. Current i_1 flows in the arm AB while i_2 in arm AD. Current i_1, on reaching the junction B gets further divided into two parts $(i_1 - i_g)$ and i_1, along branches BC and BD respectively. At junction D, currents i_2 and i_g are added to give a current $(i_2 + i_g)$ in the branch DC. Finally, currents $(i_1 - i_g)$ and $(i_2 + i_g)$ add up at junction C to give a current $(i_1 + i_2)$ or i along branch CE. In this way, currents are distributed in the different branches of bridge. In this position, we get a deflection in the galvanometer.

(iii) Now the resistances P, Q, R and S are so adjusted that on pressing the key K_2, deflection in the galvanometer becomes zero or current i_g in the branch BD becomes zero. In this situation, the bridge is said to be balanced.

(iv) In this balanced position of bridge, same current i_1 flows in arms AB and BC and similarly same current i_2 in arms AD and DC. In other words, resistances P and Q and similarly R and S, will now be joined in series.

(v) *Condition of Balance* : Applying Kirchoff's 2nd law to mesh ABDA,

$$i_1P + i_gG - i_2R = 0 \quad ...(1)$$

Similarly, for the closed mesh BCDB, we get

$$(i_1 - i_g) Q - (i_2 + i_g) S - i_gG = 0 \quad ...(2)$$

When bridge is balanced, $i_g = 0$.

Hence eq. (1) & (2) reduce to

$$i_1 P - i_2 R = 0 \text{ or } i_1 P = i_2 R \quad ...(3)$$

$$i_1 Q - i_2 S = 0 \text{ or } i_1 Q = i_2 S \quad ...(4)$$

Dividing (3) by (4), we have

$$\frac{P}{Q} = \frac{R}{S} \quad ...(5)$$

This is called as condition of balance for wheatstone's Bridge.

(vi) It is clear from above equation that if ratio of the resistances P and Q, and the resistance R are known, then unknown resistance S can be determined. This is the reason that arms P and Q are called as Ratio arms, arm AD as known arm and arm CD as unknown arm.

(vii) When the bridge is balanced then on interchanging the positions of the galanometer and the cell there is no effect on the balance of the bridge. Hence, the arms BD and AC are called as conjugate arms of the bridge.

(viii) The sensitivity of the bridge depends upon the value of the resistances. *The sensitivity of bridge is maximum when all the four resistances are of the same order.*

Potentiometer

(i) *It is an accurate device for measuring the emf of a cell or the potential difference between two points of an electric circuit.*

(ii) It consists of a very long (4 - 12 meters) and uniform wire which is made of a material having high specific resistance and low temperature coefficient of resistance such as constantan, nichrome etc. The wire is spread on a wooden board in the form of parallel wires, each of 1 meter length.

(iii) A cell C, whose e.m.f. is to be measured, is also connected with the wire in such a way that its +ve terminal is connected with one end P of the wire and –ve terminal is connected through a galvanometer G to a jockey J which is free to slide along the wire and can be made to touch it at any point.

(iv) When current from the battery flows through the wire from end P to end Q, a potential difference is produced across the wire. The fall of potential per unit length of the wire is called as Potential Gradient (= K).

(v) (a) *When jockey is pressed on some point L on extreme LHS,* then potential difference between point Pand L < e.m.f. of cell (= E_c).

Since potential at point P > potential at L, hence current from battery B (shown by → and represented by i_b) will flow through galvanometer along the path PCL. But because the +ve terminal of the cell C is also connected to the point P, hence current from cell (shown by ⇒ and represented by i_c) will flow through the galanometer along the path LCP i.e. two currents i_b, and i_c will be in opposite directions. But because e.m.f. of the cell is greater than potential difference between point P and L, hence $i_c > i_b$ i.e. resultant current will flow through the galvanometer in the direction LCP and needle of the galvanometer is deflected towards one side.

(b) *When jockey is pressed at some point R on extreme RHS,* then

Potential difference between point P and R > e.m.f. of cell E_c. and $i_b > i_c$. In this situation, resultant current will flow through the galvanometer in the direction PLC and the needle of the galvanometer is deflected towards opposite side.

(c) *When jockey is pressed at some point N in between the points L and R :* It is now evident that in between points L and R there will be a point, say N, such that when jockey is made to touch it, there will be no defection in galvanometer. This point N is called as Null point. In this situation. *Potential difference between point A and N will be equal to the e.m.f. of the cell.*

Note : Because in Null position, there is no current in the cell circuit, hence cell will be effectively in an open circuit and potential difference across its terminals will be equal to its e.m.f.

(vi) If i is the current flowing through the wire then in null position then, $E_c = V_{PN} = ir_{PN} = (i\ \rho\ l)/A$

where r_{PN} = resistance of the portion PN of the wire

l = length of portion PN or balance length

ρ = Resistivity of the material of the wire

A = Cross sectional area of the wire

Hence $(E_c/l) = (\rho/A)$ ...(1)

If R and L represent the resistance and length of the complete potentiometer wire then

$$R = \rho\left(\frac{L}{A}\right)$$

or $\frac{\rho}{A} = \frac{R}{L}$ = Resistance per unit length

If cross sectional area of the wire is uniform, then its resistance per unit length will be constant at constant temperature.

$$\text{Hence } \frac{E_c}{l} = \frac{\text{Current} \times \text{Resistance of wire}}{\text{length of wire}}$$

$$= \frac{\text{Potential difference across the wire}}{\text{Length of the wire}}$$

= Potential gradient = K i.e. ($E_c = K\,l$)

(viii) **Important Points**

(a) Higher is the length of potentiometer wire, lesser is the potential gradient along the wire and therefore higher is the balance length even for a very small e.m.f. which can be measured more accurately.

(b) The wire must be uniform otherwise K will be different at different points of the wire and readings will be inaccurate.

(c) E_b must be greater than E_c otherwise e.m.f. will not be balanced even over the complete length of wire.

(d) +ve terminals of both the battery and cell must be connected at same point P otherwise i_b and i_c will be in same direction and null point is never obtained.

IMPORTANT LINE

1. For a given, thermocouple, the variation of thermo emf. E with temperature of hot junction θ is given by $E = \alpha\,\theta + \frac{1}{2}\,\beta\,\theta^2$, where α and β are constants for a thermocouple.

   Thermoelectric power or Seebeck coefficient,

$$S = \frac{dE}{d\theta} = \alpha + \beta\theta$$

2. The graph between S and θ is a straight line.
3. Seebeck effect, Peltier effect and Thomson's effect are reversible effects.
4. Temperature of inversion depends upon the temperature of cold junction whereas the neutral temperature is independent of temperature of cold junction of a thermocouple. The value of neutral temperature is constant for a thermocouple but not so for temperature of inversion.
5. Termoelectric power in a thermocouple is zero at neutral temperature.

   If, $E = \alpha\,\theta + \frac{1}{2}\,\beta\theta^2$, then $S = \frac{dE}{d\theta} = \alpha + \beta\,\theta$

At neutral temperature, $\frac{dE}{d\theta} = 0$,

so $0 = \alpha + \beta\,\theta_n$ or $\theta_n = -\frac{\alpha}{\beta}$.

If temp. of cold junction $\theta_0 = 0°C$ then

$$\theta_i = 2\,\theta_n = -2\alpha/\beta.\ \text{So}\ \theta_i = 2\theta_n.$$

6. The amount of heat evolved or absorbed at a junction on passing the current, according to Peltier effect is

$$H = \pi\, I\, t$$

where π is the Peltier coefficient.

7. For Peltier effect or Thomson's effect, the heat evolved or absorbed is directly proportional to current. But for Joule law of heating effect, the heat produced is directly proportional to the square of the current flowing through it.

8. If S, π and σ are the Seebeck coefficient, Peltier coefficient and Thomson's coefficient respectively, then it is found that

(i) $S = \frac{dE}{dT} = \frac{\pi}{T}$

(ii) $\sigma = -T\frac{d^2E}{dT^2}$

$$= -T\frac{d}{dT}\left(\frac{dE}{dT}\right)$$

$$= -\frac{TdS}{dT}$$

9. Seebeck effect is the resultant of Peltier effect and Thomson's effect.

10. Thomson's coefficient of lead is zero.

11. Law of successive metals;

$$E_A^D = E_A^B + E_B^C + E_C^D.$$

12. Law of successive temperatures;

$$E_{T_1}^{T_n} = E_{T_1}^{T_2} + E_{T_2}^{T_3} + \ldots + E_{n-1}^{T_n}$$

13. Joule's law of heating. It states that the amount of heat produced in a conductor is directly proportional to the:

(i) square of the current flowing through the conductor,

(ii) resistance of the conductor and

(iii) time for which the current is passed.

14. *Electric power.* It is defined as the rate at which work is done in maintaining the current in electric circuit.

    Electric power, $P = VI = I^2R = V^2/R$ watt or joule/second.

15. *Electric energy.* The electric energy consumed in a circuit is defined as the total work done in maintaining the current in an electric circuit for a given time

    $$\text{Electric energy} = VIt$$
    $$= Pt = I^2 Rt$$
    $$= V^2t/R$$

    S.I. unit of electric energy is joule (denoted by J) where

    $$1 \text{ joule} = 1 \text{ watt} \times 1 \text{ second}$$
    $$= 1 \text{ volt} \times 1 \text{ ampere} \times 1 \text{ second}$$

    Commercial unit of electric energy is kilowatt hour (k Wh)

    where $1 \text{ kWh} = 1000 \text{ Wh}$

    $$= 3.6 \times 10^6 \text{ J}$$

16. *Electrolysis.* The process of decomposition of a solution into ions on passing the current through it is called electrolysis.

17. *Electrolyte.* It is a substance which allows the current to pass through it and also decomposes into positive and negative ions. For example, acids, bases, salts dissolved in water, alcohol are common electrolytes. AgI is a solid state electrolyte and KCl, NaCl are electrolyte in their molten state.

18. *Electrodes.* These are the two metal plates which are partially dipped in the solution for passing the current through the electrolyte.

19. Ni-Fe cell has lower efficiency than lead acid accumutator and its internal resistance is more than that.

20. *Anode.* The electrode connected to the positive terminal of the battery i.e. the electrode at higher potential, is called anode.

21. Cathode. The electrode connected to the negative terminal of the battery i.e. the electrode at lower potential, is called cathode.

22. Ions. The charged constituents of the electrolyte which are liberated on passing current are called ions.

23. Anions. The ions which carry negative charge and move towards the anode during electrolysis are called anions.

24. Cations. The ions which carry positive charge and move towards the cathode during electrolysis are called cations.

25. Voltameter. The vessel, containing electrodes and electrolyte, in which the electrolysis is carried out is called a voltameter.

26. In copper voltameter. The electrodes are of copper plates and electrolyte is an aqueous solution of $CuSO_4$. During electrolysis copper is removed from the anode and is deposited at the cathode.

27. Seebeck effect. It is the phenomenon of generation of an electric current in a thermocouple by keeping its two junctions at different temperatures.

    Seebeck found that the magnitude and direction of thermo e.m.f. developed in a thermocouple depends upon (i) the nature of metals forming a thermo couple and (ii) difference in temperature of the two junctions. Seebeck effect is a reversible effect. It means, if hot and cold junctions are inter-changed, the direction of thermo electric current is reversed.

28. Thermocouple. The assembly of two different metals joined at their ends to have two junctions in a circuit, is called a thermocouple.

29. Seebeck series. Seebeck, from his experimental investigation, arranged a number of metals in a series known as Seebeck series. Some of the metals, in this series, in the order Seebeck arranged them are bismuth, nickel, platinum, silver, gold, copper, lead, zinc, iron and antimony.

30. Direction of thermo electric current in copper-iron thermocouple is from copper to iron through hot junction. It can be recollected by word chi. In Sb-Bi, thermocouple, the direction of thermo electric current is from Sb to Bi through cold junction (This can be recollected by the words ABC.)

31. Current is defined as the rate of flow of charge. It is a scalar quantity. S.I. unit of current is ampere.

32. Current flowing normally through unit area of cross section is called current density. It is a vector quantity it is represented by J, $i = \int \vec{J} \cdot \vec{ds}$.
33. The distance between two successive collisions is called free path and the average of all the free paths is called the mean free path.
34. Time taken between two successive collisions is called relaxation time. It is of the order of 10^{-14} sec.
35. The electric field gives a constant velocity to the free electrons along the length of the wire. This velocity is called drift. Velocity (Vd) of electrons. The order of drift velocity is 10^{-4} m/sec.
36. For making connection wires we must used materials having low specific resistance so that resistance of connection wires may be minimum possible and they may not effect the current in a circuit in which they are used.
37. Relation between electric current and drift velocity

    $q = (nA\ V_d t)e.$

    where

    n = no. of electrons per unit volume

    A = cross sectional area of the conductor

    V_d = drift velocity of electron

    t = time taken by electron to drift l distance.

    e = charge of electron.

    $$I = neAV_d.$$
38. Resistance of a conductor is directly proportional to the length of the conductor and inversely proportional to the cross sectional area R α l and $R \propto \frac{l}{A}$

    or $R \propto \frac{l}{A}$ or $R = \rho \frac{l}{A}$

    where ρ is the specific resistance. The S.I. unit of resistance is ohm. Where $\rho = \frac{m}{ne^2\tau}$.

39. Conductance is the reciprocal of the resistance $G = \frac{1}{R}$.

40. When temperature of a metal wire increases the root mean square velocity of electron increases and relaxation time (τ) decreases. By which resistance and resistivity of the conductor, increases.

41. Resistivity of alloys increases with rise in temperature but this increase is much smaller as compared to pure metals. Because of high resistivity and negligible temperature coefficient of resistance. These alloys are used for standard resistance and resistance boxes etc.

42. Those substances whose electric conductance is large as compared to insulators but small as compared to conductors. These are called semiconductors.

43. The resistivity of semiconductors decreases with rise in temperature. i.e. temperature coefficient of resistance is negative.

44. The reason for decrease in resistance with rise in temperature of semi conductor by relation $R = \frac{ml}{ne^2\tau A}$ the value of n goes on increasing due to the breakage of covalent bonds.

45. The resistivity of electrolytes also decreases with rise in temperature. The reason is that with rise in temperature the viscosity of electrolytes decreases so that ions get more freedom to move inside the electrolytes. Hence the resistivity of the electrolytes decreases.

46. 1 ampere current is = 6.25×10^{18} electrons/sec.

47. The temperature coefficient of thermistors is negative and it is high.

48. 1 Killo watt hour = 36×10^5 Joules.

49. Killowatt hour is the unit of consumed electrical energy.

50. If two bulbs of power P_1 and P_2 are connected in parallel and the rated voltage is applied, then the total power consumed is $P = P_1 + P_2$.

51. If two bulbs of Power P_1 and P_2 are connected in series and the rated voltage is developed across each bulb, then the total power consumed is $\frac{1}{P} = \frac{1}{P_1} + \frac{1}{P_2}$.
52. For same materials $\propto$ (coefficient of thermal resistance) is nearly zero. There is no change in resistance with temperature.
53. If α is positive the resistance increase with increase in temperature.
54. If α is negative the resistance decreases with increase in temperature.
55. In parallel resistance circuit, the potential difference across each resistor is equal to the applied potential difference and the current through each resistor is inversely proportional to the resistance of that resistor.

$$I_1 : I_2 : I_3 = \frac{1}{R_1} : \frac{1}{R_2} : \frac{1}{R_3}$$

or $I_1 R_1 = I_2 R_2 = I_3 R_3$ = a constant.

56. Effective resistance R in parallel circuit.

$$\frac{1}{R} = \frac{1}{R_1} + \frac{1}{R_2} + \frac{1}{R_3}$$

Effective conductance,

$$G = G_1 + G_2 + G_3.$$

57. For n equal resistances

$$\frac{R_{series}}{R_{parallel}} = \frac{nR}{R/n} = n^2.$$

58. In parallel combination of resistances the potential difference across all the resistances is the same.
59. The current through each branch is inversely proportional to the resistance of that branch in parallel circuit.
60. As impurity in a resistance increases the resistivity of a conductor.
61. If the radius of the metallic wire becomes n times its resistance becomes $(1/n^4)$ times.
62. If length of metallic wire is made n times then its resistance becomes n^2 times but resistivity of conductor remains unchanged.

63. Using n conductors of equal resistances the number of combinations one can have using all at a time is $2^n - 1$.
64. If the resistances of n conductors are entirely different, then the number of possible combination are 2^n.
65. Drift velocity of electrons in a conductor is given by $V_d = \frac{I}{neA}$.

    Where n is the no. of electrons per unit volume of the conductor.
66. $\propto$ is positive for pure metals and alloys but negative for semiconductors and electrolytes.
67. Circuits involving Diodes, Triodes, Transistors etc. are non-ohmic circuits.
68. Heat produced in a conductor

$$H = \frac{I^2Rt}{4.2} = \frac{V^2t}{4.2\,R} = \frac{Vit}{4.2}.$$

69. Rate of production of heat in a conductor or resistance

    $P = i^2R = \frac{V^2}{R} = Vi.$
70. Mechanical stress in a conductor increases the resistivity.
71. Fuse wire in a circuit is used to control the maximum current flowing in a circuit. It is a thin wire having high resistance and is made up of a material with low melting point.
72. In metals the charge carriers are free electrons.
73. In liquids the charge carriers constituting currents are positive and negative ions.
74. In gases the charge carriers are positive ions and free electrons.
75. The direction of current density is the direction of motion of positive charge at that point.
76. When a wire is folded n times on its own length to $(1/n)^{th}$ of its length then the new resistance becomes R/n^2 where R is the initial resistance of the wire.
77. Magnetic field increases the resistivity of all metals.
78. Resistance of a conductor increases with decreases in density or when it is subjected to mechanical stress.

79. The conductor behaves as super conductors at a very low temperature.
80. The value of temperature coefficient of resistance of super conductor is zero.
81. Ferromagnetic material eg. iron, nickel and cobalt. Whose resistivity decreases with the applied magnetic field.
82. When current is drawn from the cell, E.M.F. of a cell = Terminal potential difference + voltage drop across the internal resistance of a cell. The direction of current inside the cell is from negative terminal to positive terminal.
83. During charging of a cell terminal potential difference = E.M.F. of a cell + voltage drop across internal resistance of a cell i.e. terminal potential difference becomes greater than the e.m.f. of the cell. The direction of current inside the cell is from +ve terminal to –ve terminal.
84. A device that can maintain a constant potential difference across its terminals is called as source of E.M.F.
85. The e.m.f. is the potential difference across the terminals of a cell when no current is drawn from it.
86. During the passage of an electric current through a cell its electrolyte also offers resistance to the flow of current, called as internal resistance.
87. The resistivity of a conductor increases in (i) impurity (ii) temperature and (iii) mechanical stress.
88. The resistivity of an alloy is greater than the resistivity of its constituents.
89. The resistivity of Antimony, Bismuth and semiconductors decreases with increase of temperature.
90. Due to internal resistance of the cell (a) the energy is consumed inside the cell in a closed circuit. (b) there is a potential drop inside the cell.
91. Kirchhoff's first law is the law of conservation of charge and Kirchhoff's second law is the law of conservation of energy.
92. Wheatstone bridge is most sensitive when the resistance in all the four arms of the bridge is of the same order.

93. The balanced position of the Wheatstone bridge is not affected on interchanging the positions of battery and galvanometer.
94. Greater the length of the potentiometer wire smaller is the potential gradient K = V/l and more is the balancing length hence more is the accuracy of observations.
95. Energy supplied by the cell in t seconds = Eit if (r = 0) or Vit if r is finite where E is the E.M.F. of the cell and V is the potential difference across the terminals of the cell.
96. Energy dissipated in the cell due to its internal resistance is = i^2rt.
97. If the diameter of the potentiometer wire is not uniform its potential gradient will not be uniform.
98. While performing the experiment of meter bridge the cell key is to be pressed first and galvanometer key later on to avoid inductive effect in the circuit.
99. When e.m.f. of a cell is measured by a potentiometer then in the position of null point no current flows in the cell circuit. i.e. the cell is in open circuit. Hence we obtain the actual value of the e.m.f. of the cell. Thus, a potentiometer is equivalent to an ideal voltmeter of infinite resistance.
100. Internal resistance of a cell is measured by potentiometer whose l_1 and l_2 are the length in open and closed circuit across a cell and R is the external resistance then the internal resistance of the cell is

$$r = R\left(\frac{l_1}{l_2} - 1\right).$$

101. Potentiometer is an ideal voltmeter, since potentiometer measures the e.m.f. in null position, there is no error involved of reading the scale.
102. Nichrome is an alloy is used as electrical heater element because of its high specific resistance, high melting point and low temperature coefficient of resistance.
103. Heater wire is always connected in parallel to the mains.
104. Post office box is based on the Wheatstone bridge principle. It is used to measure the unknown resistance of a wire.
105. Meter bridge is more sensitive than a post office box.

106. The draw back of meter bridge is the appearance of end resistance and the effect of this end resistance is reduced by interchanging the gaps.

107. The resistance of potentiometer can be considered as infinity while measuring the e.m.f.

108. Maximum power is delivered to load only when the internal resistance of the source is equal to the load resistance (R) then $P_{max} = \frac{V^2}{4R}$.

109. Efficiency of a source of current is $\eta = \frac{R}{R+R} \times 100$ output power is maximum if R = r then $\eta = \frac{R}{R+R} \times 100 = 50\%$.

110. Electrochemical equivalent of a substance is defined as the ratio of equivalent weight to the Faraday constant.

111. Chemical equivalent is defined as the ratio of atomic weight to the valency of the element.

112. From Faraday's first law of electrolysis

$$m = Zq = ZIt.$$

where Z is the electrochemical equivalent of the electrolyte.

113. If ρ is the density of the material deposited and A is the area of deposition, then the thickness (d) of the layer deposited in electroplating process is $d = \frac{m}{\rho A} = \frac{ZIt}{\rho A}$.

114. One Faraday = 96500 C/gm. mol.

115. 96500 coulomb charge is required to liberate 1 mole of Hydrogen.

116. Electrochemical equivalent of a substance = Electro chemical equivalent of hydrogen × chemical equivalent of the substance.

117. $Z = \frac{1}{Ne} \frac{M}{p} = \frac{E}{F}$ ($\because$ F = Ne)

where N is the Avogadro's number

= 6.023×10^{23} atoms/gm. mol.

M is the atomic mass of the substance in gram and e is the charge of electron

$$e = 1.6 \times 10^{-19} \text{ coulomb}.$$

Z = Electrochemical equivalent of the element.

118. Brightness of a bulb is directly proportional to the rate of production of heat of the bulb.

119. Resistance of an electric bulb varies inversely as its power.

120. Heat, lost per second per unit surface area of a fuse wire, $H = \frac{I^2\rho}{2\pi^2 r^3}$ or $I^2 \propto r^3$ or $I \propto r^{3/2}$, which is independent of the length of the fuse wire.

121. If t_1 and t_2 are the time taken by two different coils for producing same heat with same supply then,

   (a) If they are connected in series to produce same heat, time taken is $t = t_1 + t_2$.

   (b) If they are connected in parallel to produce same heat time taken is

$$\frac{1}{t} = \frac{1}{t_1} + \frac{1}{t_2} \text{ or } t = \frac{t_1 t_2}{t_1 + t_2}.$$

122. For a given thermo couple, the variation of thermo e.m.f. E with temperature of hot function θ is given by $E = \alpha\theta + \frac{1}{2}\beta\theta^2$ where α and β are constants for a thermo couple. Thermoelectric power or seebeck coefficient,

$$S = \frac{dE}{d\theta} = \alpha + \beta\,\theta.$$

123. Seebeck effect, Peltier effect and Thomson's effect are reversible effects.

124. For Peltier effect or Thomson's effect, the heat evolved or absorbed is directly proportional to current. But the Joule law of heating effect, the heat produced is directly proportional to the square of the current flowing through it.

125. Temperature of inversion depends upon the temperature of cold function where as the neutral temperature is independent of temperatures of cold junction of a thermo couple.

126. The amount of heat evolved or absorbed at a function on passing the current, according to Peltier effect is $H = \pi It$ where π is the Peltier coefficient.

127. The seebeck coefficient (s), Peltier coefficient (π) and Thomson coefficient (σ) is given by

$$S = \frac{dE}{dT} = \frac{\pi}{T}$$

$$\text{(ii)}\ \sigma = -T\frac{d^2E}{dT^2} = -T\frac{d}{dT}\left(\frac{dE}{dT}\right) = -\frac{TdS}{dT}.$$

128. Seebeck effect is the resultant of Peltier effect and Thomson's effect.

129. Ni-Fe coil has lower efficiency than lead acid accumulator and its internal resistance is more than that.

130. Thomson's coefficient of lead is zero.

131. Effect of stretching a wire on its resistance

There are three cases.

(a) If the length of the wire is changed, its mass remains constant. Let l and r be the length and radius of the wire and d be the density of the wire, then mass of the wire, $m = \pi r^2 ld$ or $\pi r^2 = m/ld$.

Resistance, $R = \rho \frac{l}{\pi r^2} = \frac{\rho l}{m/ld} = \frac{\rho l^2 d}{m}$ i.e. $R \propto l^2$

It means if the length of the wire is stretched to n times (i.e. $l_1 = nl$) then new resistance R_1 will be

$$\frac{R_1}{R} = \left(\frac{nl}{l}\right)^2 = n^2$$

or $$R_1 = n^2 R$$

(b) If the radius of the wire is changed, its mass remains constant.

So, $m = \pi r^2\, l\, d$ or $l = \frac{m}{\pi r^2 d}$

Now, $R = \frac{\rho l}{\pi r^2} = \frac{\rho}{\pi r^2} \times \frac{m}{\pi r^2 d} = \frac{\rho m}{\pi^2 r^4 d}$ i.e. $R \propto \frac{1}{r^4}$

It means if radius of the wire is made n times (i.e. $r_1 = nr$), then new resistance of the wire R_1 will be

$$\frac{R_1}{R} = \frac{r^4}{(nr^2)} = \frac{1}{n^4} \quad \text{or} \quad R_1 = \frac{R}{n^4}$$

(c) If area of cross-section of the wire is changed, its mass m remains constant, so

$$m = Al d \quad \text{or} \quad l = m/Ad$$

$\therefore$ Resistance, $R = \frac{\rho l}{A} = \frac{\rho}{A} \times \frac{m}{A\,d} = \frac{\rho m}{A^2 d}$ i.e. α $1/A^2$

If the area of cross-section of the metallic wire is made n times, (i.e., $A_1 = nA$) then its resistance R_1 becomes

$$\frac{R_1}{R} = \frac{A^2}{A_1^2} = \frac{A^2}{(nA)^2} = \frac{1}{n^2}$$

or $R_1 = R/n^2$

132. In series resistance circuit

(i) The current in each resistance is same but voltage across different resistances is different.

(ii) The total resistance in the circuit is the sum of the individual resistances, including internal resistance of the cell (if any), which is more than the maximum individual value of resistance in the circuit i.e.

$$R = R_1 + R_2 + R_3 + \ldots.$$

(iii) The total potential diff. across the series combination of resistances is equal to sum of potential difference across individual resistance i.e.

$$V = V_1 + V_2 + V_3 + \ldots\ldots.$$

(iv) The effective conductance G of the circuit is

$$\frac{1}{G} = \frac{1}{G_1} + \frac{1}{G_2} + \frac{1}{G_3} + \ldots$$

where g/2. g/2. G_3 ... are the conductances of the various resistances in series of the circuit.

133. In parallel resistance circuit

(i) The potential difference across each resistance is the same but current through different resistance is different.

(ii) The reciprocal of total resistance in the circuit is the sum of the reciprocal of the individual resistances i.e.

$$\frac{1}{R} = \frac{1}{R_1} + \frac{1}{R_2} + \frac{1}{R_3} + \ldots$$

(iii) The effective conductance G of the circuit is given by G $= G_1 + G_2 + G_3 + ...$

where g/2. g/2. G_3 are the conductances of the various resistors in parallel combination of a circuit.

(iv) The effective resistance is less than the smallest resistance in the combination.

134. The resistance of a conductor increases with decrease in density or when it is subjected to some mechanical stress.

135. If the value of temperature coefficient of resistance of a material is zero (i.e. $\alpha = 0$), there will be no change in its resistance with temperature. The temperature coefficient for super-conductor is zero.

136. The resistivity of a conductor increases with increases in (i) impurity (ii) temperature and (iii) mechanical stress.

137. The resistivity of an alloy is greater than the resistivity of its constituents.

138. The resistivity of Antimony, Bismuth and semiconductors decreases with increases of temperature.

139. Magnetic field increases the resistivity of all metals except iron, cobalt and nickel (which are ferromagnetic materials). The magnetic field decreases the resistivity of ferromagnetic materials like iron, cobalt and nickel.

140. Closed circuit means that current is drawn from the cell.

141. Due to internal resistance of the cell:

(a) The energy is consumed inside the cell in a closed circuit.

(b) There is a potential drop inside the cell.

142. When current is drawn from the cell, the potential difference across the terminals of the cell is less than the e.m.f. of the cell (E) by an amount equal to potential drop across the internal resistance of the cell i.e. $V = E - Ir$. V goes on decreasing on taking more and more current from cell.

143. When the cell is charged, the potential difference (V) across the terminals of the cell is greater than the e.m.f. of the cell (E) by an amount equal to potential drop across the internal resistance of cell i.e. $V = E + Ir$.

Charging current

$$= \frac{\text{e.m.f. of charger} - \text{e.m.f. of a cell}}{\text{total resistance of circuit}}$$

115. In a battery of number of cells, one wrongly connected cell destroys one more cell. Therefore in a battery of N cells each of e.m.f. E if n cells are wrongly connected, then the effective e.m.f. of the battery is E' = NE – 2nE.

116. When external resistance R is connected to a cell of e.m.f. E and internal resistance r, the power transferred from the cell to the resistance is maximum only if r = R.

118. The flow of electrons from A to B, will make the conventional current from B to A.

119. The slope of the graph showing the variation of potential difference (V) on X-axis and current (I) on Y-axis gives the conductance of the conductor carrying current.

120. The reciprocal of slope of V – I graph gives the resistance.

121. The resistance as well as conductance of a conductor depend upon (i) the length (ii) the area of cross-section, (iii) nature of material of conductor and (iv) temperature of the conductor.

122. Free electron density (n) in a metal is given by n = N x d/M, where N = Avogadro's number; x is the no. of free electron per atom; d is the density of metal and M is the atomic weight of metal.

123. If length of metallic wire is made n times then its resistance becomes n^2 times but resistivity of conductor remains unchanged.

124. If the radius of the metallic wire becomes n times, its resistance becomes $(1/n^4)$ times.

125. If the area of cross-section of the metallic wire becomes n times, then its resistance becomes $1/n^2$ times.

126. Resistance of a conductor increases with decrease in density or when it is subjected to mechanical stress.

127. Resistance of pure metals and metallic alloy increases with increase in temperature but the resistance of semiconductor decreases with increase in temperature.

128. The value of temperature coefficient of resistance (a) of a conductor is different at different temperature. The temperature coefficient of resistance averaged over the temperature range t_1°C to t_2°C is given by $\alpha = \frac{R_2 - R_1}{R_1(t_2 - t_1)}$, where R_1, R_2 = resistance of conductor at t_1°C and t_2°C respectively.

129. The conductor behaves as superconductor at a very low temperature.

130. The value of temperature coefficient of resistance of superconductor is zero.

131. In series resistance circuit, the current is same in every resistor and the voltage across any resistor is directly proportional to the resistance of that resistor.

i.e. $V_1 : V_2 : V_3 = R_1 : R_2 : R_3$

or $\frac{V_1}{R_1} = \frac{V_2}{R_2} = \frac{V_3}{R_3}$ = constant.

Effective resistance,

$$R = R_1 + R_2 + R_3$$

Effective conductance,

$$\frac{1}{G} = \frac{1}{G_1} + \frac{1}{G_2} + \frac{1}{G_3}$$

132. In parallel resistance circuit, the potential difference across each resistor is equal to the applied potential difference and the current through each resistor is inversely proportional to the resistance of that resistor.

i.e. $I_1 : I_2 : I_3 = \frac{1}{R_1} + \frac{1}{R_2} + \frac{1}{R_3}$

or $I_1R_1 = I_2R_2 = I_3R_3$ = a constant.

Effective resistance R is;

$$\frac{1}{R} = \frac{1}{R_1} + \frac{1}{R_2} + \frac{1}{R_3}$$

Effective conductance,

$$G = G_1 + G_2 + G_3$$

133. For n equal resistances;

$$\frac{R_{series}}{R_{parallel}} = \frac{nR}{R/n} = n^2$$

134. Using n conductors of equal resistances, the maximum number of combination one can have, using all at a time is $2^n - 1$.

135. Using n conductors of different resistances, the number of possible combinations are 2^n.

136. Resistivity of a conductor increases with
   (i) increase in impurity in conductor
   (ii) increase in temperature of conductor
   (iii) increase in mechanical stress on conductor.

137. Magnetic field increases the resistivity of all metals excepts iron, cobalt and nickel (i.e. ferromagnetic material). The resistivity of iron, cobalt and nickel decreases with the magnetic field applied.

138. When current is drawn from the cell (i.e. during discharging of a cell), e.m.f. of a cell = terminal potential difference + voltage drop across the internal resistance of a cell. The direction of current inside the cell is from negative terminal to positive terminal.

   During charging of a cell, Terminal potential difference = e.m.f. of a cell + voltage drop across internal resistance of a cell i.e. terminal potential difference becomes greater than the e.m.f. of the cell. The direction of current inside the cell is from + ive terminal to – ve terminal.

139. In series grouping of cells, the effective e.m.f. on n cells, each of e.m.f. E becomes nE. In parallel grouping of n cells each of e.m.f. E, the effective e.m.f. is E because in parallel combination of cells, the sizes of the electrodes will increase without affecting the e.m.f. of the cell.

140. The current in the external resistor will be maximum
   (i) in series grouping of cells, provided the value of internal resistance of a cell is very very small as compared to external resistance,
   (ii) in parallel grouping of cell, provided the value of internal resistance of a cell is very very large as compared to external resistance,

(iii) in mixed grouping of cells, provided the value of external resistance is equal to total internal resistance of all the cells.

141. In a battery of number of cells if one cell is wrongly concerted, it will destroy the e.m.f. of one more cell. In a battery of N cells, each of e.m.f. E, if n cells are wrongly connected, then the e.m.f. of the battery is E' = N E – 2 n E.

142. E.C.E. of any substance = E.C.E. of hydrogen × its chemical equivalent.

143. The potentiometer is equivalent to an ideal voltmeter of infinite resistance because while measuring the e.m.f. of a cell by a potentiometer, at the position of null point, no current flows in the cell circuit, i.e. the cell is in the open circuit. Hence we obtain actual value of e.m.f. of the cell.

144. In metre bridge, the balance point on bridge wire is not effected on interchanging the positions of battery and galvanometer.

145. The effect of end resistances in metre bridge can be reduced by interchanging the resistances in the two gaps and repeating the experiment for the observation in the middle of the wire.

146. The sensitiveness of potentiometer can be increased by increasing the potential gradient along the potentiometer wire.

147. Current Carriers. The charged particles whose flow in a definite direction constitutes the electric current are called current carriers. e.g. electrons in conductors, ions in electrolyte, electrons and holes in semiconductor.

148. Electromotive force of a cell is defined as the maximum potential difference between the two electrodes of a cell when the cell is in the open circuit. E.M.F. of a cell depends upon nature of electrodes, nature and concentration of electrolyte used in the cell and its temperature.

149. Electric Current. The flow of charge in a definite direction constitutes the electric current.

$$\text{Electric current} = \frac{\text{charge flowing}}{\text{time taken}} = \frac{q}{t}$$

S.I. Unit of current is ampere (denoted by A). The arrow head marked in circuit represents the direction of conventional current i.e. direction of flow of positive charge, whereas the direction of flow of electrons gives the direction of electronic current which is opposite to that of conventional current. Current is a scalar quantity.

150. Drift Velocity : If is defined as the average velocity with which free electrons get drifted towards the positive end of the conductor under the influence of an external electric field. Drift velocity of electrons is given by:

$$\vec{v}_d = -\frac{e\vec{E}}{m} t$$

where e is the charge on electron, m is the mass, $\vec{E}$ is the electric field applied and τ is the time of relaxation i.e. the average time that has elapsed since each electron suffered its last collision with the ion or atom of the conductor, while drifting towards the positive end of the conductor under the effect of external electric field applied.

The value of drift velocity of electron is about 10^{-5} m/s and value of relaxation time is about 10^{-14} second.

151. Relation between current and drift velocity.

$$I = nAev_d$$

where n is the electron density or no. of electrons per unit volume of the conductor and A is the area of cross-section of the conductor. The small value of the drift velocity ($\sim 10^{-5}$ ms^{-1}) produces a large amount of current due to presence of large number of free electrons in a conductor ($\sim 10^{29}$ m^{-3}). The propagation of current is at the speed of light.

152. Ohm's Law : It states that the current (I) flowing through a conductor is directly proportional to the potential difference (V) across the ends of the conductor, provided physical conditions of the conductor such as temperature, mechanical strain etc. are kept constant i.e.

$$V \propto I \text{ or } V = IR$$

where R is known as resistance of the conductor, which depends upon the nature and dimensions of the conductor. The S.I. unit of R is ohm.

153. Resistance of a conductor. It is the obstruction posed by the conductor to the flow of current through it. Resistance of a conductor is due to the collisions of free electrons with the ions or atoms of the conductor while drifting towards the positive

end of the conductor. The resistance of a conductor can be given by the expression:

$$R = \frac{m}{ne^2\tau}\frac{l}{A} = \rho\frac{l}{A}$$

where m is the mass of electron, e is charge of electron, n is the number density of electron, τ is the relaxation time, l is the length of conductor and A is its area of cross section, ρ is the specific resistance of the conductor.

154. Specific resistance or electrical resistivity of the material of a conductor is defined as the resistance of unit length and unit area of cross-section of the conductor. The S.I. unit of resistivity is Ωm. Resistivity of a conductor depends upon the nature of the conductor but is independent of the length or area of cross section of the conductor. In fact resistivity,

$$\rho = \frac{m}{ne^2\tau}$$

155. Current density (J) at a point is defined as the amount of current flowing per unit area of cross-section of the conductor, provided the area is held in a direction normal to the current.

$$J = \frac{i}{A} = nev_d$$

The S.I. unit of current density is Am^{-2}

156. Electrical conductivity (σ) of a conductor is the inverse of its resistivity (ρ) i.e. $\sigma = 1/\rho$. The S.I. unit of σ is $\Omega^{-1}\ m^{-1}$ or $S\ m^{-1}$

157. Effect of temperature on resistance. The resistance of a metal conductor at a temperature t°C is given by $R_t = R_0\ (1 + \alpha t)$ where R_0 is the resistance of a conductor at 0°C and α is the temperature co-efficient of resistance.

For metals α is positive i.e. resistance increases with rise in temperature.

For semi conductors and insulators α is negative i.e. resistance decreases with rise in temperature.

For alloys like manganin, eureka and constantan, the value of α is very small as compared to that of conductors. That is why these alloys are used in making standard resistance.

If R_{t_1} and R_{t_2} are the resistances of the same conductor at temperature t_1°C and t_2°C, then $R_{t_2} = R_{t_1} [1 + \alpha (t_2 - t_1)]$. Here α is the temperature coeff. of resistance averaged over the temperature range t_1°C and t_2°C.

158. Non-ohmic conductors. Those conductors which do not obey Ohm's law are called non-ohmic conductors e.g. vacuum tube, liquid electrolyte etc.

159. Super-conductors. Those materials which offer least resistance to the flow of current through them are called super-conductors. Examples : mercury at temp 4.2 K, lead at 7.25 K and niobium at temperature 9.2 K become super-conductors.

160. Colour code for carbon resistors. The number attached from 0 to 9 to the various colours can be recollected by the sentence B.B. ROY Great Britain Very Good Wife.

Black – 0, Brown – 1, Red – 2, Orange – 3, Yellow – 4, Green – 5, Blue – 6, Violet – 7, Grey – 8, White – 9. The strip of gold, silver and no colour shows the accuracy of 5%, 10% and 20% of the given carbon resistor.

161. Chemical equivalents of hydrogen, copper and silver are 1.008, 31.75 and 108 respectively.

162. 96500 C are required to liberate 1.008 g of hydrogen at cathode in water voltameter during electrolysis.

163. 2.016 g of hydrogen occupies 22.4 litres at S.T.P.

164. Resistance in series. The total resistance (R_s) is given by

$$R_s = R_1 + R_2 + R_3 + \ldots$$

165. Resistance in parallel. The total resistance (R_p) is given by

$$\frac{1}{R_p} = \frac{1}{R_1} + \frac{1}{R_2} + \frac{1}{R_3} + \ldots$$

166. Internal resistance of a cell is defined as the resistance offered by the electrolyte and electrodes of a cell when electric current flows through it. Internal resistance of a cell depends upon : (i) distance between the electrodes, (ii) the nature of electrodes (iii) nature of electrolyte and (iv) area of the electrodes immersed in the electrolyte.

167. Terminal potential difference of a cell is defined as the potential difference between the two electrodes of a cell in a closed circuit. Terminal potential difference of a cell decreases if the current drawn from the cell increases. Terminal potential difference of a cell (V) is less than the e.m.f. of a cell (E) by an amount equal to potential drop across the internal resistance of the cell i.e.

$$V = E - Ir$$

$$\text{or } r = \frac{E - V}{I} = \left(\frac{E - V}{V}\right) R \quad (\because V = IR)$$

where R is the external resistance in the circuit and r is the internal resistance of a cell.

168. Kirchhoff's Laws

First law. The algebraic sum of the current meeting at a junction is zero i.e. $\sum i = 0$. The current reaching a function if taken positive then the current leaving the junction is taken negative. This law supports the concept that moving charges are not accumulated at a junction.

Second law. In a closed loop, the algebraic sum of the emfs is equal to the algebraic sum of the products of the resistance and the respective currents flowing through them i.e.

$$\sum E = \sum IR.$$

While traversing a closed loop (in clockwise or anti clock wise direction), if negative pole of the cell is encountered first then its emf is positive, otherwise negative. The product of resistance and current in an arm of the circuit is taken positive if the direction of current in that arm is in the same sense as one moves in a closed loop and is taken negative if the direction of current in that arm is opposite to the sense as one moves in the closed loop.

169. Principle of potentiometer. It is based on the fact that the fall of potential across any portion of the wire is directly proportional to the length of that portion provided the wire is of uniform area of cross section and a constant current is flowing through it. i.e.

$$V \propto l \text{ (if I and A are constant)}$$

or $\quad V = Kl$

where K is called potential gradient i.e. fall of potential per unit length of the given wire.

170. Expression for comparison of emfs. of two cells by using potentiometer, $\frac{E_1}{E_2} = \frac{l_1}{l_2}$

where l_1, l_2 are the balancing lengths of potentiometer wire for the emfs E_1 and E_2 of two cells.

171. Expression for the determination of internal resistance of a cell (r) by potentiometer method

$$r = \left(\frac{l_1 - l_2}{l_2}\right)R$$

where, l_1 = balancing length of potentiometer wire corresponding to e.m.f. of the cell.

l_2 = balancing length of potentiometer wire corresponding to terminal potential difference of the cell when a resistance R is connected in series with the cell whose e.m.f. is to be determined.

172. When current I is passed through a resistor of resistance R for time t, then heat produced is

$$H = I^2 Rt \text{ joule} = \frac{I^2 Rt}{4.2} \text{ calories.}$$

173. Due to Joule-heating effect, power consumed by a resistor R in watts will be

$$P = \frac{H}{t} = I^2R = VI = \frac{V^2}{R} \qquad [\because V = IR]$$

174. If resistances (or electrical appliances) are connected in series, the current through each resistance is same. Then power of an electrical appliance

$P \propto R$ and $P \propto V \qquad (\because V = IR)$

It means in series combination of resistance, the potential difference and power consumed will be more in larger resistance.

175. If resistances (i.e. electrical appliances) are connected in parallel, the potential difference across each resistance is same. Then

$P \propto I/R$ and $I \propto I/R$ [as $V = IR$]

It means in parallel combination of resistances the current and power consumed will be more in smaller resistances.

176. For a given voltage V, if resistances is charged from R to (R/n), power consumed changes from P to nP.

$P = V^2/R$; when $R' = R/n$,

then $P' = V^2/(R/n) = nV^2/R = nP$.

177. (i) When the appliances of power P_1, P_2, P_3 ... are in series, the effective power consumed (P) is

$$\frac{1}{P} = \frac{1}{P_1} + \frac{1}{P_2} + \frac{1}{P_3} + \ldots$$

(ii) If n appliances, each of equal resistance R, are connected in series with a voltage source V, the power dissipated P_s will be

$$P_s = \frac{V^2}{nR} \qquad \ldots(i)$$

(iii) When the appliances of power P_1, P_2, P_3 ... are in parallel, the effective power consumed (P) is

$$P = P_1 + P_2 + P_3 + \ldots$$

(iv) If n appliances, each of equal resistance R, are connected in parallel with a voltage source V, the power dissipated P_s will be

$$P_P = \frac{V^2}{(R/n)} = \frac{nV^2}{R} \qquad \ldots(ii)$$

from (i) and (ii), $\frac{P_P}{P_S} \frac{P_P}{P_S} = n^2$

or $P_P = n^2 P_S$.

It means power consumed by n equal resistances in parallel is n^2 times of power consumed in series if voltage remains same.

178. In parallel grouping of bulbs across a given sources of voltage, the bulb of greater wattage will give more brightness and will allow more current through it, but will have lesser resistance and same potential difference across it.

179. In series grouping of bulbs across a given source of voltage, the bulb of greater wattage will give less bright light and will

have lesser resistance and potential difference across it but same current.

180. If I is the current through the fuse wire of length l, radius r, specific resistance ρ, resistance R and h is the rate of loss of heat per unit area of a fuse wire, then at steady-state

$$I^2 R = h\,A \text{ or } \frac{I^2 \rho l}{\pi r^2} = h \times 2\pi r l$$

or $\quad I \propto r^{3/2}$

It shows that the current capacity of a fuse is independent of its length and varies with its radius as $r^{3/2}$.

181. If t_1 and t_2 are the time taken by two different coils for producing same heat with same supply, then

(i) if they are connected in series to produce same heat, time taken $t = t_1 + t_2$

(ii) if they are connected in parallel to produce same heat, time taken is, $t = \dfrac{t_1 t_2}{t_1 + t_2}$.

182. Maximum power is delivered to the load only when the internal resistance of the source is equal to the lad resistance (R). Then $P_{max} = \dfrac{V^2}{4R}$.

183. Efficiency of a source of current is

$$\eta = \frac{R}{R+r} \times 100$$

output power is maximum, if R = r, then

$$h = \frac{R}{R+R} \times 100 = 50\%$$

184. Kirchhoff 's first law is the law of conservation of charge and Kirchhoff's second law is the law of conservation of energy.

185. Wheatstone bridge is most sensitive when the resistance in all the four arms of the bridge is of the same order.

186. The balanced position of the Wheatstone bridge is not affected on interchanging the positions of battery and galvanometer.

187. In the balanced condition of the bridge, the current through galvanometer arm BD is zero so potential of B = potential of D.

188. The resistance between A and C of Wheatstone bridge in balanced position is

$$R_{AC} = \frac{(P+Q)(R+S)}{(P+Q+R+S)}$$

189. While performing the experiment of metre bridge, the cell key is to be pressed first and galvanometer key later on to avid inductive effect in the circuit.

190. The drawback of meter bridge is the appearance of end resistances which can be reduced by (i) taking the readings in middle of bridge wire and (ii) by interchanging the resistances of gaps and repeating the experiment.

191. The sensitiveness of a potentiometer (i.e. smallest potential difference that can be measured with its help) can be increased by decreasing its potential gradient along the potentiometer wire. The same can be achieved.

(i) by increasing the length of potentiometer wire or (ii) by reducing the current in the potentiometer wire from the main battery with the help of rheostat, it the potentiometer wire is of fixed length.

192. Potentiometer is an ideal voltmeter.

193. E.C.E. if a substance,

$$z = \frac{\text{equivalent weight}}{\text{Faraday}} = \frac{E}{F}$$

where 1 F = 96500 $C/_{\text{gm mol.}}$ and

$$E = \frac{M}{P} = \frac{\text{atomic weight}}{\text{valence}}$$

194. From Faraday's first law of electrolysis $m = zq = zIt = z\left(\frac{P}{V}\right)t$.

195. If ρ is the density of the material deposited and A is the area of deposition, then the thickness (d) of the layer deposited in electroplating process is $d = \frac{m}{\rho A} = \frac{zIt}{\rho A}$.

196. The back e.m.f. for water voltameter is 1.67 and it is 1.34 V for $CuCl_2$ electrolytes voltameter with platinum electrodes.

197. Equivalent weight of hydrogen, Cu and Ag are 1.008, 31.5 and 108 respectively.

198. 96500 C are required to liberate 1.008 g of hydrogen.

199. 2.016 g of hydrogen occupies 22.4 litres at N.T.P.

200. E.C.E. of a substance = E.C.E. of hydrogen × chemical equivalent of the substance.

201. V-I curve for a voltameter is a straight line beyond the voltage of back e.m.f. of electrolyte, hence ohm's law is obeyed there.

202. Seebeck effect is a reversible effect.

203. The variation of therm e.m.f. with temperature of hot junction when cold junction of a thermocouple is kept at low temperature is a parabolic curve.

204. Neutral temperature is independent of temperature of cold junction whereas temperature of inversion depends upon the temperature of cold junction.

205. The variation of thermo electric power with temperature of hot junction of a thermocouple is a straight line.

206. (i) The termoelectric power is positive if temperature of hot junction lies in between temp. of cold junction and neutral temperature.

    (ii) The thermoelectric power is negative if temperature of hot junction lies in between neutral temperature and temperature of inversion.

207. Thermoelectric power is minimum at neutral temperature of a thermocouple.

208. Peltier effect and themson's effect are reversible effects.

209. Lead shows zero Thomson's effect.

210. Peltier coefficient (π) of a junction is $\pi = T\dfrac{dE}{dT} = T \times S$ where S is the Seebeck coefficient.

211. Thomson's coefficient,

$$\sigma = -T\frac{d^2E}{dT^2} = -T\frac{dE}{dT}\left(\frac{dE}{dT}\right) = -T\frac{dS}{dT}$$

212. Law of intermediate metals $E_A^C = E_B^C + E_A^B$.

213. Law of intermediate temperature

$$E_{T_1^n}^T = E_{T_{n-1}^n}^T + E_{T_{n-2}^{n-1}}^T + \dots + E_{T_1^2}^T.$$

214. Resistance of an electric bulb varies inversely as its power i.e. $R \propto 1/P$.

215. When the various electrical appliances are connected in series, their power decreases. The effective power P_s in series combination is given by

$$\frac{1}{P_S} = \frac{1}{P_1} + \frac{1}{P_2} + \frac{1}{P_3} + \dots$$

216. When the various electrical appliances are connected in parallel, their power increases. The effective power P_P in parallel combination is given by

$$P_P = P_1 + P_2 + P_3 + \dots$$

217. Brightness of a bulb is directly proportional to the rate of production of heat of the bulb.

218. In houses, all the electrical appliances are connected in parallel. The voltage across each appliance is the same but current through them depends upon the power of the appliance. The higher power appliance draws more current and lower power appliance draws less current. ($\because I = P/V$ or $I \propto P$)

219. In parallel combination of electric bulbs, the resistance of 100 W bulb is less than that of 50 W bulb.

220. In series combination of electric bulbs, the brightness of 50 W bulb is more than that of 100 watt bulb.

221. If t_1 and t_2 are the time taken by two different coils for producing same heat with same supply, then

    (i) if they are connected in series to produce same heat, time taken is $t = t_1 + t_2$

    (ii) if they are connected in parallel to produce same heat, time taken is $\frac{1}{t} = \frac{1}{t_1} + \frac{1}{t_2}$

    or $t = \frac{t_1 t_2}{t_1 + t_2}$

222. When the bulbs of different wattages but of same rated voltages are connected in series and if the voltage supplied is twice the rated voltage, then the bulb with the least wattage fuses off.

223. Heat lost per second per unit surface area of a fuse wire,

$$H = \frac{I^2\rho}{2\pi^2 r^3} \quad \text{or} \quad I^2 \propto r^3$$

or $I \propto r^{3/2}$; which is independent of the length of the fuse wire.

224. If E is the e.m.f. of a cell and r is the internal resistance of a cell, then the maximum current that can be drawn from the cell is, $I_{max} = E/r$.

225. The power dissipated in the external resistance in a circuit by the cell is maximum when the value of external resistance (R) is equal to the internal resistance (r) i.e. R = r.

226. When the power dissipated in the external resistance is maximum, the efficiency of the cell is 50%.

227. In general, the efficiency of a cell is given by

$$\eta = R/(R + r)$$

228. The energy is dissipated in the cell due to its internal resistance.

229. The power input or the electric power supplied by the cell is P_i = E.M.F. × current = EI

$$= E^2/(R + r)$$

230. The power dissipated in the external circuit is

$$P_0 = VI$$

where V = Pot. diff. across load resistance and I is the current through it.

231. The charging current I_c for a secondary cell is given by

$$I_c = \frac{\text{emf of charger} - \text{eimf of cell}}{\text{total resistance of the circuit}}$$

232. The chemical equivalent expressed in grams is called gram equivalent.

233. The charge required to liberte one gram equivalent of the substance is called faraday. The value of faraday constant

= 96500 C/gm. equivalent.

SOLVED EXAMPLES

Example 1:

A plate of area 10 cm^2 is to be electroplated with copper (density = 9 gms/cm^3) to a thickness of 0.001 cm on both sides using a battery of 12 volts. Calculate the energy spent by the cell in the process of deposition. If this energy is used to heat 100 gms of water, calculate the rise in temperature of the water, ECE of copper = 0.0003 gms/coul.

Solution :

Total surface area of the plate which is to be electroplated

$= A = 20 \text{ cm}^2.$

Hence, mass of copper deposited

m = volume × density = Area × thickness × density

$= 20 \times .001 \times 9 = 0.18$ gm

According to faraday's 1st law : $m = Zit$

$$\text{or, } it = \frac{m}{Z} = \frac{0.18}{0.0003} = 600 \text{ Coulombs}$$

As $V = 12$ volts, hence energy spent by the cell

$= Vit = 12 \times 600 =$ *7200 Joules*

If $d\theta$ is the rise in temperature of the water,then

$$ms\, d\theta = \frac{7200}{4.2} \text{ cal}$$

$$\text{or } 100 \times 1 \times d\,\theta = \frac{7200}{4.2},$$

$$\therefore d\theta = \frac{7200}{4.2 \times 100} = \mathbf{17.14\,^\circ C}$$

Example 2:

A piece of metal weighing 200 gms is to be electroplated with 5% of its weight in gold. If the strength of the available current is 2 amperes, how long would it take to deposit the required amount of gold, E.C.E. of H = 0.1044 × 10^{-4} g./C, atomic weight of gold = 197.1 and atomic weight of hydrogen = 1.008.

Solution :

Mass of gold to be deposited, m = [200 × (5/100)] = 10 gms

i = 2 amp, W_2 = chemical equivalent of gold = (197.1/3) = 65.7

W_1 = chemical equivalent of Hydrogen = 1.008

Hence,according to Faraday's 2nd law :

$W_1/Z_1 = W_2/Z_2$.

$\therefore$ W.C.E. of gold = $Z_2 = (W_2/W_1) \times Z_1 = \dfrac{65.7}{1.008} \times 0.1044 \times 10^{-4}$ g/C

From Faraday's first law : m = Zit

$$\text{or } t = \frac{m}{Zi} = \frac{10}{(65.7/1.008) \times 0.1044 \times 10^{-4} \times 2}$$

$$= \frac{10 \times 1.008}{65.7 \times 0.1044 \times 10^{-4} \times 2}$$

= 7347.9 seconds = **2 hours 2 min 27.9 sec.**

Example 3:

A 3 ton motor operated vehical goes up an inclined plane having a slope of sin^{-1} *(1/25), at an uniform speed of 36 km/hour. The tractive resistance may be taken as 40 kgf per ton. Assuming effciency of a motor = 85% and mechanical efficiency between motor and road wheels = 80%, calculate*

(i) output of motor, and

(ii) current taken by the motor which works on 240 volts.

Solution :

(i) m = mass of vehicle = 3 × 1000 = 3000 kg

Speed of motor = v = 36 km/hour = 36 × (5/18) = 10 m/sec.

Tractive resistance = F = 3 × 40 = 120 kgf = 120 × 9.8 = 1176 N.

Component of the weight of the car along the plane

= Wsin θ = 3000 × 9.8 × (1/25) = 1176 N.

As the vehicle is moving upwards, hence tractive resistance will also act down the plane.

Hence total force down the plane = 1176 + 1176 = 2352 N.

∴ Total power output at road wheels

= Total Force × speed = 2352 × 10 = 23520 Watt.

Because this is 80% of motors ' output, we have output of motor

= 23520/0.8 = 29400 Watt

(ii) Because efficiency of motor is 85%, hence power input from mains

$$= \frac{29400}{0.85} = 34588 \text{ Watt.}$$

i.e. Vi = 34588,

$$\therefore \quad i = \frac{34588}{V} = \frac{34588}{240} = \mathbf{144\ Amp.}$$

Example 4:

Two heater coils made of the same material are connected in parallel across the mains. The length and diameter of the wire of one of the coils are double that of the other. Which one of them will produce more heat?

Solution :

We know that

$$P = \frac{V^2}{R} \text{ and } R = \frac{\rho L}{A},$$

hence, $P = \dfrac{V^2 A}{\rho L}$

Since potential difference applied is same across both wires, hence

$(PL/A) = (V^2/\rho)$ = constant

$$\text{or}, \quad \frac{P_1 L_1}{A_1} = \frac{P_2 L_2}{A_2}$$

$$\text{or}, \quad \frac{P_1 L_1}{\pi r_1^2} = \frac{P_2 L_2}{\pi r_2^2}$$

$$\text{or}, \quad \frac{P_2}{P_1} = \frac{L_1}{L_2} \times \frac{r_2^2}{r_1^2} = \frac{L_1}{2L_2}\left(\frac{2r_1}{r_1}\right)^2 = 2$$

or $P_2 = 2P_1$.

i.e. power devoloped in 2nd wire = 2 × power developed in 1st wire

i.e. *2nd wire will develop more heat as compared to that developed by 1st wire in the same time.*

Example 5:

A copper wire having cross sectional area of 0.5 mm^2 and length of 0.1 m is initially at 25° C and is thermally insulated from its surroundings. If a current of 10 amps is set up in the wire-

(i) Find the time in which the wire will start melting. The change of resistance with the temperature of the wire may be neglected.

(ii) What will be the time if length of wire is doubled ? Given for copper, density = 9×10^3 kg/m^3.

Specific heat = 9×10^{-2} Kcal/kg- °C, melting point = 1075° C, specific resistance = 1.6×10^{-8} Ω -m.

Solution :

(i) If l is the length of the wire, A is the cross sectional area and ρ is the specific resistance of the material then resistance R of the wire is given by $R = \rho l/A$

If i is the current flowing in the wire, then electric energy supplied in t seconds, for increasing the temperature of the wire upto melting point, is given by

$$H = \frac{i^2Rt}{4.2} = \frac{i^2\rho l t}{A \times 4.2} \text{ cal} \qquad ...(1)$$

Mass of the wire = m = l Ad

Heat energy required for increasing the temperature upto melting point, is given by

$$H = m\,s\,(\theta_2 - \theta_1) = l\,A\,d\,s\,(\theta_2 - \theta_1) \qquad ...(2)$$

Hence from equations (1) & (2)

$$\frac{i^2\rho l t}{A \times 4.2} = l\,A\,d\,s\,(\theta_2 - \theta_1)\,;$$

$$t = \frac{A^2 ds\,(\theta_2 - \theta_1) \times 4.2}{i^2\rho} \qquad ...(3)$$

$$\text{or} \quad t = \frac{(5\times10^{-7})^2\times9\times10^3\times9\times10^1\times4.2\times1050}{(10)^2\times1.6\times10^{-8}}$$

= **558.14 second.**

(ii) Because t is independent of the length of the wire, hence there will be no change in time if length of wire is doubled.

Example 6:

A heater using a Nichrome element operates at 220 volts. When it is switched on at 0° C, it carries an initial current of 2.75 ampere. After few seconds, the current reaches a steady value of 2.44 amp. What is the final temperature of the element? The average value of the temperature coefficient of the resistance of Nichrome over the temperature range is 0.0045/° C.

Solution :

Given that at 0° C, V = 220 volts, i = 2.75 amp

Hence resistance of the wire at 0° C

R_0 = (V/i) = (220 / 2.75) = 80 ohms

If final temperature is t°C, then resistance of the wire at t°C

R_t = (220/2.44) = 90.16 ohms;

Now, $R_t = R_0\,(1 + \alpha t)$

$$\therefore \quad t = \frac{R_1 - R_0}{R_0\times\alpha} = \frac{90.16-80}{80\times45\times10^{-5}} = \mathbf{282^\circ C}$$

Example 7:

The density of platinum is 21.4 gm/cm³ and its resistivity is 1.0 × 10⁻⁷ ohm-m. Assume the wire to be rectangular with dimension of 0.01 cm and 0.02 cm. If the mass of the wire is 51.36 mg, what is the resistance of the wire ?

Solution :

If L is the length and A the cross sectional area of the wire, then

$$L\,A = \text{Volume of wire} = \frac{\text{mass}}{\text{density}} = \frac{m}{d},$$

$$\therefore \ L = \frac{m}{Ad}$$

Hence Resistance of the wire is given by

$$R = \rho\frac{L}{A} = \rho\frac{m}{A^2 d}$$

$$= \frac{10^{-7} \times 51.36 \times 10^{-6}}{(2 \times 10^{-8})^2 \times 21.4 \times 10^3} = \mathbf{0.6\ ohms}$$

Example 8:

A 200 V DC generator transmits power to a distant place through cables of total length 5 km. If the cable has resistance of 0.08 ohm / km, what is the efficiency of transmission when the cable current is 150 A ? How is the efficiency altered by transmitting the same power at 500V?

Solution :

(i) Potential difference across the cable = current × resistance of the cable

= 150 amp × (.08 ohms/km) × 5 km = 60 volt

Output potential = 200 – 60 = 140 volt

∴ Efficiency of transmission

$$= \frac{\text{Output voltage}}{\text{Input voltage}} \times 100 = \frac{140 \text{ Volt}}{200 \text{ Volt}} \times 100$$

= **70%**

(ii) Delivered power = Vi = 200 × 150 = 30,000 Watts, Generated emf = 500 volt

It i is the current transmitted, then Voltage drop across the cable

= i × 0.4 = **0.4 i Volt**

Since transmitted power is same in both cases, the current

$$= \frac{200 \times 150}{500} = 60 \text{ amp}$$

∴ Voltage drop across cable

= 0.4 × 60 = 24 volt

$$\text{Hence efficiency} = = \frac{500 - 24}{500} \times 100$$

$$= \frac{476 \times 100}{500} = \mathbf{95.2\%}$$

Example 9:

Two metallic conductors, one made of copper and other of iron are connected in parallel across some source of constant e.m.f. At 20° C, currents flowing in both are equal. If temperature is raised to 100° C, what proportion of current will pass through each conductor. Given that α for Cu = 0.0042/° C and α for Iron = 0.006/° C.

Solution :

Because currents flowing in both the conductors are equal at 20° C, hence their resistance are equal at 20° C.

If R is the resistance of each conductor at 20°C, then resistance of copper conductor at 100° C.

$R_1 = R\,[1 + \alpha_1 \times 80] = R\,[1 + .0042 \times 80]$

= 1.336 R and that of iron conductor at 100°C

$R_2 = R\,[1 + \alpha_2 \times 80] = R\,[1 + .006 \times 80]$

= 1.48 R

Because R_1 and R_2 are joined in parallel hence currents through them will be inversely proportional to their resistance i.e.

$$\frac{i_1}{i_2} = \frac{R_2}{R_1} = \frac{1.48\ R}{1.336\ R} = \frac{1480}{1336}$$

or, $$\frac{i_1 + i_2}{i_2} = \frac{1480 + 1336}{1336} = \frac{2816}{1336},$$

$$\therefore \quad \frac{i_2}{i_1 + i_2} = \frac{1336}{2816}$$

i.e. percentage proportion of i_2

= (1336/2816) × 100 = **47.44%.**

∴ percentage proportion of i_1 = **52.56%**

Example 10:

Two heaters A and B are connected in parallel across the suply voltage V. Heater A generates 500 × 10³ calories of heat in 20 minutes while B generates 1000 × 10³ calories in10 minutes. The resistance of heater A is 10 ohms. What will be the resistance of heater B ? If these heaters are connected in series across the same voltage, how much heat will be produced in 5 minutes ?

Solution :

Because two heaters are connected in parallel across the suply voltage V hence voltage across each = V. Hence heat produced by two heaters are given by

$$H_1 = \frac{V^2 t_1}{4.2 R_1} \text{ and } H_2 = \frac{V^2 t_2}{4.2 R_2} \qquad ...(1)$$

$$\therefore \frac{H_1}{H_2} = \frac{R_2}{R_1} \times \frac{t_1}{t_2}$$

$$\text{or } \frac{500 \times 10^3}{1000 \times 10^3} = \frac{R_2}{R_1} \times \frac{20}{10}$$

As R_1 = 10 ohms, hence $R_2 = (R_1/4)$ = **2.5 ohms**

when two heaters are connected in series, their total resistance

$$= R_1 + R_2 = 12.5 \text{ ohms}$$

$\therefore$ Heat produced by combination

$$H = \frac{V^2}{4.2\ R} t \text{ calories} = \frac{V^2 \times 300}{4.2 \times 12.5} \text{ calories} \qquad ...(2)$$

From equation (1),

$$H_1 = \frac{V^2 t_1}{4.2 R_1} \qquad ...(3)$$

Dividing eq. (2) by eq. (3),

$$\frac{H}{H_1} = \frac{300}{t_1} \times \frac{R_1}{12.5}$$

$$\therefore \quad H = \frac{300}{t_1} \times \frac{R_1}{12.5} \times H_1$$

$$= \frac{300 \times 10 \times 500 \times 10^3}{20 \times 60 \times 12.5} = \mathbf{100 \times 10^3 \text{ calories}}$$

Example 11:

A wire of length 1.0 m and radius 10 ^{3} m is carrying a heavy current and is a assumed to radiate as a black body. At equilibrium its temperature *is 900 K white that of the surroundings is 300 K. The resistivity of the material of the wire at 300 K is $\pi^2 \times 10^{-8}$ Ω m and its temperature coefficient of resistance is 7.8×10^{-3} per ^{o}C. Find the current in the wire. (Given : Stefan Constant = 5.68×10^{-8} w m^{-2} K^{-4})*

Solution :

Given that length of the wire l = 1.0 m, radius of the wire r = 10^{-3} m, equilibrium temperature $T_2 = 900^oK$, Temperature of surroundings $T_1 = 300^oK$, ρ_0 = Resistivity at 300°K = $\pi^2 \times 10^{-8}$ Ω m,

$\alpha = 7.8 \times 10^{-3}/^oC$

and $\sigma = 5.68 \times 10^{-8}$ Wm^{-2}K^{-4}.

Let A = Curved surface area of the wire through which energy is radiated out = $2\pi rl$.

A' = Cross sectional area of the wire = πr^2.

In equilibrium,

$$\left[\begin{array}{l}\text{Energy lost/sec, due to radiations} \\ \text{emitted by the wire}\end{array}\right] = \left[\begin{array}{l}\text{Energy produced per sec due} \\ \text{to flow of Current}\end{array}\right]$$

According to Stefan's Law, energy lost per sec through curred surface area

$$= \sigma A (T_2^4 - T_1^4) = \sigma 2\pi r l (T_2^4 - T_1^4)$$

According to Joule's law of heating, energy produced per sec due to flow of current

$$= I^2 R = I^2 \rho \frac{l}{A'} = I^2 \rho \frac{l}{\pi r^2} = I^2 \rho_0 (1 + \alpha d\theta) \frac{l}{\pi r^2}$$

Hence, $\sigma\, 2\pi r l (T_2^4 - T_1^4) = I^2 \rho_0 (1 + \alpha d\theta)(l/\pi r^2)$

$$\text{or} \quad I^2 = \frac{\sigma 2\pi r l (T_2^4 - T_1^4) \pi r^2}{\rho_0 (1 + \alpha d\theta) l}$$

$$= \frac{\sigma \times 2\pi^2 r^3 (T_2^4 - T_1^4)}{\rho_0 (1 + \alpha d\theta)}$$

$$= \frac{(5.68 \times 10^{-8}) \times 2\pi^2 \times (10^{-3})^3 \times (900^4 - 300^4)}{(\pi^2 \times 10^{-8})[1 + (7.8 \times 10^{-3}) \times (900 - 300)]}$$

$\therefore$ **I = 35.5 amp.**

Example 12:

A charged capacitor of 5×10^{-2} F capacity is discharged through a resistor R of 20 Ω and a copper voltmeter of internal resistance 30 Ω connected in series. If 4.62×10^{-6} kg copper is deposited, calculate the heat generated in the resistor R.(ECE of copper is 3.3×10^{-7} kg/C)

Solution :

If charge on the plates of capacitor is q, then from Faraday's law of electrolysis

$$m = Zq$$

$$\text{or } q = \frac{m}{Z} = \frac{4.62 \times 10^{-6}}{3.3 \times 10^{-7}} = 14 \text{ Coul}$$

Hence energy stored in the capacitor

$$U = \frac{q^2}{2C} = \frac{(14)^2}{2 \times 5 \times 10^{-2}} = 1960 \text{ Joule}$$

In series the current i is same and heat energy produced, $H = i^2Rt$

or $H \propto R$

$\therefore$ Heat generated in the resistor R

$$H_1 = \frac{20}{20+30} \times \text{total Energy (U)} = \frac{20}{50} \times 1960$$

= **784 Joule.**

Example 13:

A uniform copper wire of mass 2.23×10^{-3} kg carries a current of 1 A when 1.7 V is applied across it. Calculate its length and area of cross-section. If the wire is uniformly stretched to double its length, calculate the new resistance. Density of copper is 8.92×10^3 kg m^{-3} and resistivity is 1.7×10^{-8} Ωm.

Solution :

$$R = \text{Resistance of wire } \frac{V}{I} = \frac{1.7}{1} = 1.7 \text{ ohms}$$

$$\text{Now, } R = \rho\frac{l}{A}$$

$$\text{or } \frac{l}{A} = \frac{R}{\rho} = \frac{1.7 \text{ ohms}}{1.7 \times 10^{-8} \text{ ohms} \times \text{meter}}$$

= **10^8/meter**

$\therefore A = 10^{-8}\, l$ ...(1)

Volume of the wire

$$= lA = \frac{m}{d} = \frac{2.23 \times 10^{-3}\text{ kg}}{8.92 \times 10^3\text{ kg/m}^3}$$

or $lA = 2.5 \times 10^{-7}\text{ m}^3$ ...(2)

From (1), and (2)

$$l^2 = \frac{2.5 \times 10^{-7}}{10^{-8}} = 25$$

or $l = 5$ m

From (1), $A = 10^{-8} \times 5 = 5 \times 10^{-8}\text{ m}^2$.

When the wire is stretched 2 times, volume remains constant i.e., $l'A' = lA$, As $l' = 2l$, hence $A' = A/2$

$$\therefore\ R' = \rho\frac{l'}{A'} = \rho\frac{2l}{A/2} = 4\left(\frac{\rho l}{A}\right) = 4R$$

$= 4 \times 1.7 =$ **6.8 ohms.**

Example 14:

1 m long metallic wire is broken into two unequal parts Pand Q. P part of the wire is uniformly extended into another wire R. Length of R is twice the length of P and the resistance of R is equal to that of Q. Find the ratio of the resistance of P and R and also the ratio of length of P and Q.

Solution :

L = 1 m, Let length of part P = x

$\therefore$ Length of part Q = L – x

P part of wire is uniformly extended to another wire R, having length double of P i.e. length of extended wire = 2x

As volume will remain constant during extension hence, $xA = 2x \times A'$ i.e. $A' = \frac{A}{2}$ = c.s. area of wire R

Now, Resistance of $R = \rho.\frac{2x}{A'} = \rho.\frac{4x}{A}$

and Resistance of $Q = \rho.\frac{(L-x)}{A}$

As, resistance of R = Resistance of Q

$\therefore \ \rho.\frac{4x}{A} = \frac{\rho(L-x)}{A}$

or $4x = L - x$

or $5x = L$

or $x = \left(\frac{L}{5}\right)$

$\therefore$ Length of $P = \left(\frac{1}{5}\right)$ m

and Length of $Q = \left(\frac{4}{5}\right)$ m

$\therefore$ Ratio of lengths of P and Q = 1 : 4

$$\frac{\text{Resistance of P.}}{\text{Resistance of R}} = \frac{\rho.\frac{x}{A}}{\rho.\frac{4x}{A}} = \frac{1}{4}$$

or **1 : 4**

Example 15:

A n type silicon sample of width 4×10^{-3} m, thickness 25×10^{-5} m and length 6×10^{-2} m carries a current of 4.8 mA when the voltage is applied across the length of the sample. What is the current density ? If the free electron density is 10^{22} m^{-3} then find how much time does it take for the electrons to travel the full length of the sample?

Solution :

Given that $b = 4 \times 10^{-3}$ m, $d = 25 \times 10^{-5}$ m

$l = 6 \times 10^{-2}$ m, i = 4.8 m amp, $n = 10^{22}/m^3$.

current density,

$$j = \frac{i}{A} = \frac{i}{bd} = \frac{4.8\times10^{-3}}{4\times10^{-3}\times25\times10^{-5}}$$

$= \mathbf{4.8 \times 10^3\ amp/m^2}$.

Now, $j = n\,ev_d$

$$\therefore\ v_d = \frac{j}{ne} = \frac{4.8\times10^3}{10^{22}\times1.6\times10^{-19}} = 3\ \text{m/s}$$

$$\therefore\ t = \frac{l}{v_d} = \frac{6\times10^{-2}}{3} = 2\times10^{-2}\ \text{sec.}$$

Example 16:

A series battery of 6 cells each of e.m.f. 2V and internal resistance 0.5 Ω is charged by a 100 V d.c. supply. What resistance should be used in the charging circuit in order to limit the charging current to 8A. Using this relation, obtain (a) the power supplied by the d.c. source, (b) the power dissipated as heat and (c) the chemical energy stored in the battery in time 15 minutes.

Solution:

Given, no. of cells $n = 6$;

E.M.F. of each cell, $E = 2V$;

internal resistance of each cell, $r = 0.5\ \Omega$;

charging voltage $V = 100$ V.

Let R be the resistance used in the series of the circuit while charging the cells. Then current in the circuit will be

$$i = \frac{V - nE}{nr + R}$$

or

$$R = \frac{V - nE}{i} - nr$$

$$= \frac{100 - 6\times2}{8} - 6\times0.5$$

$$= 11 - 3 = 8\ \Omega$$

(a) Power supplied by d.c. source

$$= V\times i = 100\times = 800\ \text{W}$$

(b) Power dissipated as heat

$$= i^2\,(R + nr)$$

$$= 8^2 (8 + 6 \times 0.5)$$

$$= 704 \text{ W}$$

(c) Rate at which the chemical energy is stored

$$= 800 - 704 = 96 \text{ watt}$$

$\therefore$ Chemical energy stored in 15 min.

$$= 96 \times 10 \times 60 = 86400 \text{ J.}$$

Example 17:

Determine the current through the battery of internal resistance 0.5Ω for the circuit shown in the fig. How much power is dissipated in 6 Ω resistance?

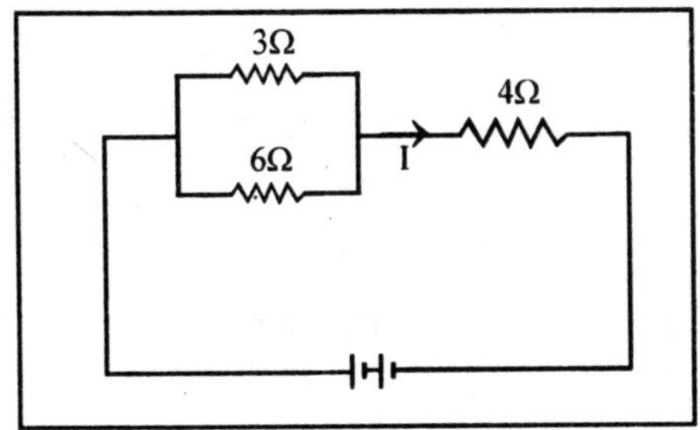

Solution:

Resistance of arm BCDE = 7 + 1 + 10 = 18 Ω. Here 18 Ω and 6 Ω are in parallel, their effective resistance,

$$R_p = \frac{18 \times 6}{18 + 6} = \frac{18 \times 6}{24} = 4.5\Omega$$

Total resistance of the circuit

$$= 2 + 4.5 + 8 + 0.5 = 15\Omega$$

$\therefore$ Current through the circuit,

$$i = \frac{15}{15} = 1 \text{ A}$$

Pot. diff. across B and E

$$= i \times R_p = 1 \times 4.5 = 4.5 \text{ V}$$

$\therefore$ Power dissipated as heat due to resistance

6 W = $(4.5)^2/6$ = 3.375 W.

Example 18:

Two uniform wires of same material each weighting 1 g but one having double the length of the other, are connected in series, carrying a current of 10A. The length of longer wire is 20 cm. Calculate the rate of consumption of energy in each of the two wires. Which one wire gets hotter? Density of the material of wire = 11 g/cc, specific resistance of the material = 20×10^{-5} Ω cm.

Solution:

Let a_1 and a_2 be the area of cross-section of shorter and longer wire respectively.

As, mass = volume × density = $al\rho$

$$\therefore \quad 1 = a_1 \times 10 \times 11 = a_2 \times 20 \times 11$$

or
$$a_1 = \frac{1}{10 \times 11} \text{ cm}^2 \text{ and } a_2 = \frac{1}{20 \times 11} \text{ cm}^2$$

$$\therefore \quad R = 20 \times 10^{-5} \times \frac{10}{1/(10 \times 11)}$$

$$= 20 \times 10^{-5} \times 10 \times 10 \times 11$$

$$= 22 \times 10^{-2}\ \Omega$$

$$\therefore \quad R_2 = 20 \times 10^{-5} \times \frac{20}{1/(20 \times 11)}$$

$$= 88 \times 10^{-2}\ \Omega$$

And rate of heat produced

$$H_1 = I^2R_1$$

$$= (10)^2 \times 22 \times 10^{-2} = 22 \text{ W}$$

and $H_2 = I^2R_2 = (10)^2 \times 88 \times 10^{-2} = 88$ W

Thus the wire of longer length gets hotter.

Example 19:

An electric coil is connected in series with a resistance of x Ω across the 240 volt, mains, the coil being immersed in 1 kg of water at 20°C. The temperature of water rises to boiling point (i.e. 100°C) in 10 minutes. When a second heating experiment is made with the resistance x short circuited, the time required to develop the same quantity of heat is reduced to 6 minutes. Calculate the value of x.

Solution:

Case (i) Let R be the resistance of heating coil. Then current through heating coil will be:

$$I = \frac{240}{X + R}.$$

$$\text{Therefore, } \left(\frac{240}{X + R}\right) R = \frac{1 \times 4.2 \times 10^3 \times (100 - 20)}{10 \times 60}$$

$$= \frac{4.2 \times 10^3 \times 80}{10 \times 60} \quad \text{...(i)}$$

Case (ii)

$$\frac{(240)^2}{R} = \frac{4.2 \times 10^3 \times (100 - 20)}{6 \times 60} \quad \text{...(ii)}$$

On solving (ii), we get

$$R = 432/7\ \Omega$$

Putting this value in (i) and solving it, we get

$$x = 17.83\ \Omega.$$

Example 20:

In a silver platinum system, an electrolysis current of 5.0 A is used for a certain time and 0.5 mole of silver is deposited. How many moles of copper and iron will be deposited in their respective plating system if an electrolysis current of 10.0 A passed for twice the time for silver plating. Relative atomic mass of silver = 107.3, of copper = 63.54, of iron = 55.85.

Solution:

$$m = zIt = \left(\frac{M}{Npe}\right) It$$

$$\text{or} \quad \frac{m}{M} = \frac{It}{Npe} \quad \text{...(i)}$$

where M = mass of a mole,

N = Avogadro's number,

p = valence and

e = charge on electron.

The value of p for silver, copper and iron is 1, 2, and 3 respectively. In case of silver m/M = 0.5

From (i), $0.5 = \frac{5 \times t}{N \times 1 \times e}$

or $\frac{t}{Ne} = \frac{0.5}{5} = 0.1$

In case of copper,

$$\frac{m}{M} = \frac{10 \times 2t}{N2e} = 10 \times 0.1$$

$$= 1 \text{ mole.}$$

In the case of Iron,

$$\frac{m}{M} = \frac{10 \times 2t}{N\,3e}$$

$$= \frac{20}{3} \times 0.1 = \frac{2}{3} \text{ mole.}$$

Example 21:

A series battery of 6 lead accumulators each of e.m.f. 2.2 V and internal resistance 0.50 Ω is charged by a 100 V d.c. supply. What series resistance should be used in the charging circuit in order to limit the current to 7.84? Using the required resistor, obtain

(a) *the power supplied by the d.c. source.*

(b) *the power dissipated as heat.*

(c) *the chemical energy stored in the battery in 15 min.*

Solution:

Total resistance

$= R + nr$

$= R + 6 \times 0.5$

$= R + 3$

Total e.m.f. of the battery

$= 6 \times 2.2 = 13.2$ V

Effective potential difference in the circuit

$= 100 - 13.2 = 86.8$ VV

$$\text{Now, current} = \frac{\text{effective pot. diff.}}{\text{total resistance}}$$

$$\therefore \qquad 7.8 = \frac{86.8}{R + 3}$$

or $\qquad R = 8.13\ \Omega$

(a) Power supplied by d.c. source

$$= VI = 100 \times 7.8 = 780 \text{ watt.}$$

(b) Power dissipated as heat

$$= I^2 (R + nr) = (7.8)^2 (8.13 + 3)$$

$$= 677.15 \text{ watt.}$$

(c) Chemical energy stored in the battery in 15 minutes.

$$= (780 - 677.15) \times 15 \times 60 = 92565 \text{ J}$$

Example 22:

A tangent galvanometer is connected in series with a copper voltameter and a steady current is passed through them. The galvanometer reading is 45° and 0.2988 gram of copper is deposited in half an hour. Find the reduction factor of tangent galvanometer. Given that E.C.E. of copper is 0.00032 gC^{-1}.

Solution:

Let I be the current in the circuit

$$\therefore \quad I = K \tan\theta = \frac{m}{zt}$$

$$\text{or} \quad K = \frac{m}{zt \tan\theta}$$

$$= \frac{0.2988}{0.000332 \times (30 \times 60) \times \tan 45^\circ} = 0.5\text{A}$$

Example 23:

A steady potential difference of 1.62 V is maintained across two platinum electrodes placed in a solution of $CuCl_2$. At the end of 600 s, the mass of copper deposited on the cathode is measured to be 5.92 g. The back e.m.f. of the voltameter is given to be 1.34 V. Estimate the *resistance of voltameter. Faraday constant = 96500 C mol ¹, relative atomic mass of copper = 63.5.*

Solution:

Here, $V = 1.62$ V;

$t = 600$ s;

$m = 5.92$ g

Back e.m.f., $V_b = 1.34$ V;

$R = ?$;

$F = 96500 \text{ C mol}^{-1}$;

Chemical equivalent of copper,

$$E = \frac{63.5}{2} = 31.75$$

As $$m = zIt = \frac{E}{F} It.$$

or $$I = \frac{mF}{Et} = \frac{5.92 \times 96500}{31.75 \times 600}$$

$$= 29.99\text{A}$$

Now, $$R = \frac{V - V_b}{I} = \frac{1.62 - 1.34}{29.99}$$

$$= 9.34 \times 10^{-3}\ \Omega$$

Example 24:

In a house having 220 V-line, the following appliances are operating: (i) 60 W bulb (ii) a 1000 W heater and (iii) a 40 W radio. Calculate (a) the current drawn by heater and (b) the current passing through fuse for this line.

Solution:

Here, $V = 200$ V;

$P_1 = 60$ W;

$P_2 = 1000$ W;

$P_3 = 40$ W.

(a) Current drawn by heater

$$= \frac{P_2}{V} = \frac{1000}{220} = \frac{50}{11} A$$

(b) Current drawn by bulb

$$= \frac{P_1}{V} = \frac{60}{220} = \frac{3}{11} A$$

Current drawn by Radio

$$= \frac{P_3}{V} = \frac{40}{220} = \frac{2}{11} A$$

Current passing through fuse for the line

$$= \frac{50}{11} + \frac{3}{11} + \frac{2}{11} = 5A.$$

Example 25:

Four resistances carrying a current as shown in Fig. are immersed in a box containing ice at 0°C. How much ice must be put in the box every 10 minutes to keep the average quantity of ice in the box constant? Latent heat of ice = 80 cal g $^{-1}$.

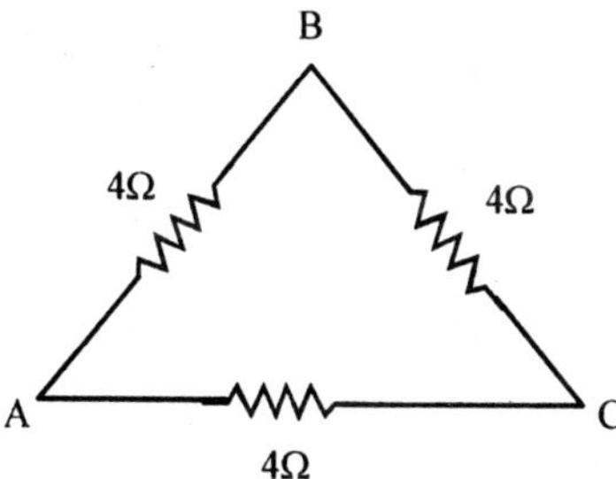

Solution:

The equivalent resistance of the circuit is

$$R = \frac{10 \times 5}{10 + 5} + \frac{5 \times 10}{5 + 10} = \frac{10}{3} + \frac{10}{3} = \frac{20}{3} \Omega$$

Heat produced in 10 min.

$$H = \frac{i^2 Rt}{J}$$

$$= \frac{(10 \times 10) \times 20 \times (10 \times 60)}{3 \times 4.2} \text{ cal}$$

Let m be the mass of the ice melted in 10 minutes

Then, $m \times 80 = \dfrac{10 \times 10 \times 20 \times 10 \times 60}{3 \times 4.2}$

or $m = \dfrac{10 \times 10 \times 20 \times 10 \times 60}{80 \times 3 \times 4.2}$

$= 1190$ gram.

Example 26:

Prove that when a current is divided between two resistances in accordance with Kirchhoff's laws, the heat produced is minimum.

Solution:

Consider two resistors R_1 and R_2 connected in parallel and the current through the various arms of the circuit be as shown in Fig. According to Kirchhoff's first law, at junction A;

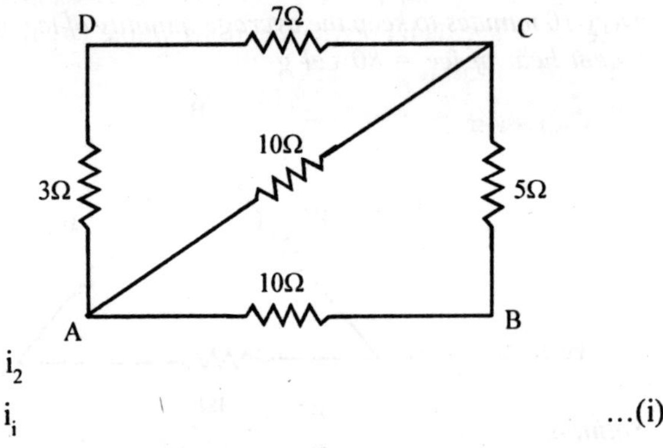

$i = i_1 + i_2$

or $i_2 = i - i_1$ ...(i)

Let H be the heat produced in the circuit in t seconds, then

$H = i_1^2 R_1 t + i_2^2 R_2 t = i_1^2 R_1 t + (i - i_1)^2 R_2 t$ [From (i)]

In case the heat produced in the circuit is minimum then $\dfrac{dH}{di_1} = 0$;

Therefore

$$2i_1R_1t + 2(i - i_1)(-1)R_2t = 0$$

or $$2i_1R_1t - 2i_2R_2t = 0$$

or $$i_1R_1 - i_2R_2 = 0$$

Which is according to Kirchhoff's second law in a closed circuit ACDEFA.

Example 27:

A series-parallel combination battery consists of 300 identical cells, each with an internal resistance 0.3 Ω. It is connected to external resistance 10 Ω. Find the number of parallel groups consisting of equal number of cells connected in series, at which the external resistance generated the higher thermal power.

Solution:

Let there be m rows of cells and n cells in each row of cells. Then total number of cells,

$$N = mn \quad \text{or} \quad n = N/m \qquad \text{...(i)}$$

Let E, r be the e.m.f. and internal resistance of each cell and R be the external resistance. Then total internal resistance of all the m rows of cells = nr/m. Total resistance of the whole circuit = (R + nr/m). Total e.m.f. of all the cells = nE.

Current in the external resistance R will be

$$i = \frac{nE}{R + nr/m} = \frac{(N/m)E}{R + (N/m)r/m}$$

$$= \frac{mNE}{m^2R + Nr}$$

Heat generated in resistance R,

$$H = i^2R = \left(\frac{mNE}{m^2R + Nr}\right)^2 R \qquad \text{...(ii)}$$

For H to be maximum,

$$\frac{dH}{dm} = 0.$$

Differentiating (ii) w.r.t. m and equating to zero, we get

$$m = \sqrt{Nr/R} = \sqrt{300 \times 0.3/10} = 3.$$

Example 28:

Power from a 64 V d.c. supply goes to charge a battery of 8 lead accumulators each of e.m.f. 2.0 V and internal resistance 1/8 Ω. The charging current also runs an electric motor placed in series with the battery. If the resistance of the windings of the motor in 7.0 Ω and the steady supply current is 3.5 A, obtain (a) the mechanical energy yielded by the motor, and (b) the chemical energy stored in the battery during charging in 1h.

Solution:

Here, Applied voltage,

$$V = 64 \text{ V}$$

Total e.m.f. of 8 lead accumulators in series,

$$E = 20 \times 8 = 16 \text{ V}$$

Total internal resistance of 8 lead accumulators,

$$r = (1/8) \times 8 = 1 \ \Omega$$

Steady current,

$$I = 3.5 \text{ A}$$

Resistance of the windings of the motor,

$$R = 7.0 \ \Omega$$

Rate of consumption of energy in the circuit

$$= VI = 64 \times 3.5 = 224 \text{ W}$$

Rate of storing chemical energy in the battery

$$= EI = 16 \times 3.5 = 56 \text{ W}$$

Rate of energy dissipation as heat in the circuit

$$= I^2 (R + r) = (3.5)^2 (7 + 1) = 98 \text{ W}$$

Rate of mechanical energy yielded by motor

$$= 224 - (56 + 98) = 70 \text{ W}$$

(a) Mechanical energy yielded by motor in 1*h*

$$= 70 \times 3600 = 252000 \text{ J}$$

(b) Chemical energy stored in the battery in 1*h*

$$= 56 \times 3600 = 201600 \text{ J.}$$

Example 29:

A copper wire having cross sectional area 0.5 mm² and a length of 0.1 m is initially at 25°C and is thermally insulated from the surroundings. If a current of 10 ampere is set up in this wire

(i) find the time in which the wire will start melting. The change of resistance with temperature of the wire may be neglected.

(ii) What will be the time taken if length of the wire is doubled?

Given for copper wire density = 9×10^3 *kg m* $^{-3}$; *specific heat* = 9×10^{-2} *k cal kg* $^{-1}$ *(°C* $^{-1}$*), melting point 1075°C, specific resistance* = 1.6×10^{-8} *ohm metre.*

Solution:

(i) Here, $\rho = 1.6 \times 10^{-8}\ \Omega$ m;

$l = 0.1$ m;

$A = 0.5 \times 10^{-6}$ m^2

$\therefore$ Resistance of copper wire, $R = \dfrac{\rho l}{A}$

$$= \frac{1.6 \times 10^{-8} \times 0.1}{0.5 \times 10^{-6}} = 0.0032\,\Omega$$

Mass of copper wire,

$$m = \text{volume} \times \text{density}$$

$$= (Al) \times \text{density}$$

$$= (0.5 \times 10^{-6} \times 0.1) \times (9 \times 10^3)$$

$$= 45 \times 10^{-5} \text{ kg}$$

sp. heat of copper, c = 0.09 k cal kg^{-1} (°C^{-1}), change in temperature.

$$d\theta = 1075 - 25 = 1050°\text{C}$$

If H is the heat required to melt the copper wire, then

$$\frac{i^2 Rt}{4.2 \times 10^3} = m \times c \times d\theta$$

or $$\frac{(10)^2 \times 0.0032 \times t}{4.2 \times 10^3}$$

$$= 45 \times 10^{-5} \times 0.09 \times 1050$$

One solving we get, t = 558 seconds.

(ii) When the length of the wire is doubled, its resistance is doubled. Now the mass of the wire is also doubled. Then

$$\frac{i^2 (2R)t}{4.2 \times 10^3} = (2\text{ m}) \times c \times d\theta$$

or $$t = \frac{mc \times d\theta \times 4.2 \times 10^3}{i^2 R}$$

It means t remains the same = 558 seconds.

Example 30(a):

An electrically heating coil is placed in a calorimeter containing 360 gram of water at 10°C. Then coil consumes energy at the rate of 90 watts. The water equivalent of the calorimeter and the coil is 40 gram. What will be the temperature of water after 10 minutes?

$J = 4.2\ J\ cal^{-1}$.

Solution:

Heat produced by heating coil

$$= \frac{90 \times 10 \times 60}{4.2} \text{ cal.}$$

Heat taken by water and calorimeter

$$= 360 \times 1 \times (\theta - 10) + 40\ (\theta - 10) \text{ cal}$$

$$\therefore \quad \frac{90 \times 10 \times 60}{4.2} = (360 + 40)\ (\theta - 10).$$

On solving we get, $\theta = 42.14°C$.

Example 30(b):

A thermocouple is made of iron and nickel and the difference of temperature between their junction is 100° C. Find the thermo e.m.f. developed across such a couple when the thermo e.m.f. for iron lead couple is 12.5 μ volt/°C and for nickel lead couple is 15μV/°C.

Solution:

Here $E_{Pb}^{Fe} = 12.5\ \mu\ V/°C$;

$E_{Ni}^{Pb} = -\ 15\mu\ V/°C$

According to law of intermediate metals,

$$E_{Ni}^{Fe} = E_{Pb}^{Fe} + E_{Ni}^{Pb} = E_{Pb}^{Fe} - E_{Pb}^{Ni}$$

$$= 12.5 - (-\ 15) = 27.5\ \mu\ V/°C.$$

So thermo e.m.f. for 100°C difference of temperature

$$= 27.5 \times 100 = 2750\ \mu\ V.$$

Example 31:

How many times will be total resistance of n identical conductors be reduced if the series arrangement is changed to parallel one.

Solution:

Effective resistance in series combination:

$$R_s = R + R + \ldots \text{ n terms} = nR.$$

Effective resistance in parallel combination

$$\frac{1}{R_P} = \frac{1}{R} + \frac{1}{R} + \ldots\ldots \text{ n terms} = \frac{n}{R}$$

or $R_P = \frac{R}{n}$ $\therefore \frac{R_p}{R_s} = \frac{R/n}{nR} = \frac{1}{n^2}$

Example 32:

The external diameter of a 5 m long hollow tube is 10 cm and the thickness of its wall is 5 mm. If the specific resistance of copper is $1.7 \times 10^{-8}\ \Omega$ m, determine its resistance.

Solution:

Here; $l = 5$ m;

$$2r_2 = 10 \text{ cm}$$

or $r_2 = 5 \text{ cm} = 0.05 \text{ m}$

$$r_1 = (5 - 0.5) \text{ cm} = 4.5 \text{ cm} = 0.045\text{m};$$

$$\rho = 1.7 \times 10^{-8}\ \Omega\text{m},$$

Here $A = p\ (r_2^2 - r_1^2)$

$$= \frac{22}{7}\ [\ (0.05)^2 - (0.045)^2]\ \text{m}^2$$

$$\therefore \quad R = \frac{\rho l}{A} = \frac{1.7 \times 10^{-8} \times 5}{(22/7)[(0.05)^2 - 0.045)^2]}$$

$$= 5.7 \times 10^{-5}\ \Omega.$$

Example 33:

The temperature coefficient of resistance of a wire is 0.00125°c^{-1}. At 300 K its resistance is one ohm. At what temperature the resistance of wire will be 2 Ω.

Solution:

Temperature 300 K = 27°C. As per question

$$R_{300} = R_0 (1 + \alpha \times 27) = 1$$

and $R_t = R_0 (1 + \alpha \times t) = 2$

$$\therefore \quad \frac{1 + 27\alpha}{1 + t\alpha} = \frac{1}{2}$$

or $2 + 54\,\alpha = 1 + t\,\alpha$

or $2 + 54 \times (0.00125) = 1 + (0.00125)\,t$

On solving, $t = 854°C = 854 + 273$

$= 1127$ K.

Example 34:

In Fig., current through 3 Ω resistance is 0.8 A, then find potential drop through 4 Ω resistance.

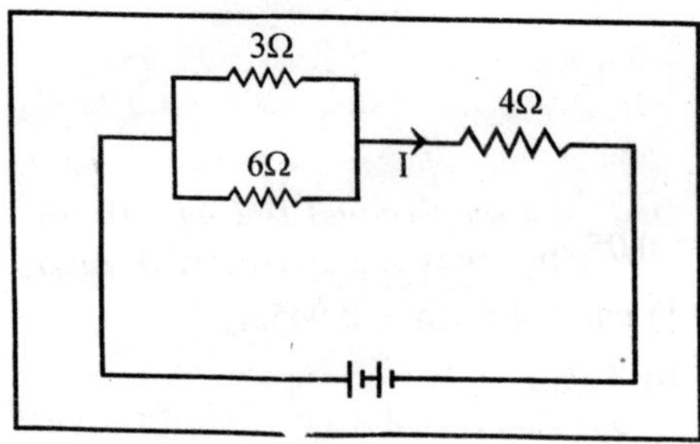

Solution:

Voltage drop across 3 Ω resistance

$= 3 \times 0.8 = 2.4$ V

This voltage drop is the same across 6 Ω resistance.

Hence current through 6 Ω resistance

$= 2.4/6 = 0.4$ A

Total current in the circuit is,

$$I = 0.8 + 0.4 = 1.2 \text{ A}$$

Voltage across 4 Ω resistance

$$= 4 \times 1.2 = 4.8 \text{ V}.$$

Example 35(a):

A 10 m long wire of resistance 20 Ω is connected in series with a battery of e.m.f. 3V (negligible internal resistance) and a resistance of 10 Ω. Find the potential gradient along the wire.

Solution:

Current, $$I = \frac{3}{10 + 20}$$

$$= \frac{1}{10} = 0.1 \text{ A}$$

Pot. diff. across 20 Ω = 0.1 × 20 = 2.0 V

Potential gradient = 2/10 = 0.2 V m^{-1}.

Example 35(b):

Twelve cells each having the same e.m.f. are connected in series and are kept in a closed box. Some of the cells are connected in reverse order. The battery is connected in series with an ammeter, an external resistance R and two cells of the same type as in the battery. The current when they aid each other is 3 ampere and is 2 ampere when the two oppose each other. How many cells are connected in reverse order?

(Roorkee 1993)

Solution:

If one cell is connected in reverse order, then it will destroy one more cell. If n cells are connected in reverse order, then 2 n cells will be destroyed. If E is the e.m.f. of each cell, then effective e.m.f. of 12 cells will be = (12 – 2n) E.

This battery if supports two similar cells in series, then supplies a current of 3 A to resistance R. Therefore,

$$3 = \frac{(12 - 2n)\,E + 2E}{R} = \frac{(14 - 2n)\,E}{R} \qquad \text{...(i)}$$

If battery and two cells oppose each other, then the current in the external resistance is 2A. Hence

$$2 = \frac{(12 - 2n)\,E - 2E}{R} = \frac{(10 - 2n)\,E}{R} \qquad \ldots\text{(ii)}$$

Dividing (i) by (ii), we get

$$\frac{3}{2} = \frac{14 - 2n}{10 - 2n} \quad \text{or} \quad n = 1$$

Example 36:

Five resistances are connected as shown in the Fig. Find the equivalent resistance between the points (i) A and B (ii) A and D and (iii) A and C.

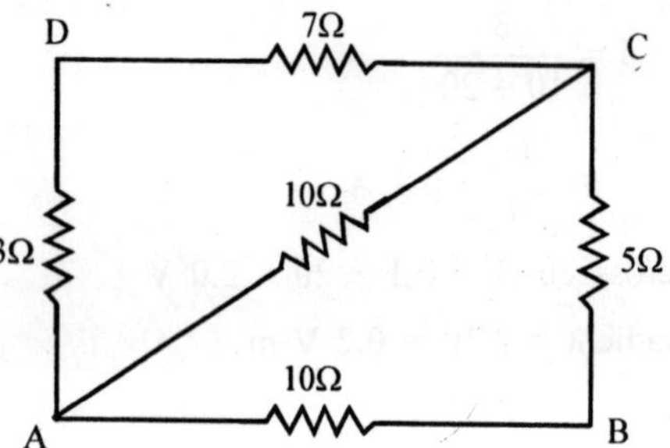

Solution:

(i) The resistance of arm ADC

$= 3 + 7 = 10\ \Omega.$

This resistance of 10 Ω is in parallel with 10 Ω resistance of arm AC. Their effective resistance

$$= \frac{10 \times 10}{10 + 10} = 5\ \Omega$$

This 5 Ω resistance is in series with 5 Ω resistance of arm CB.

Their effective resistance = 5 + 5 = 10 Ω. This 10 Ω resistance is in parallel with 10 Ω resistance of arm AB. Thus the effective resistance between A and B.

$$= \frac{10 \times 10}{10 + 10} = 5\ \Omega.$$

(ii) For the equivalent resistance between A and D, the resistance of arm ABC = 10 + 5 = 15 Ω is in parallel with the resistance of arm AC. Their effective resistance

$$= \frac{10 \times 15}{10 + 15} = 6\ \Omega.$$

This resistance is in series with 7 Ω of arm CD. Their effective resistance = 6 + 7 = 13 Ω. This resistance is in parallel with 3 Ω. Hence equivalent resistance between A and D

$$= \frac{13 \times 3}{13 + 3} = \frac{39}{16}\ \Omega.$$

(iii) For the equivalent resistance between A and C, the resistance of arms ACD, AC and ABC are in parallel. Their effective resistance R_p will be

$$\frac{1}{R_p} = \frac{1}{\cdot 3 + 7} + \frac{1}{10} + \frac{1}{10 + 5} = \frac{4}{15}$$

or $\quad R_p = \frac{15}{4}\ \Omega.$

Example 37:

What will be the reading in the ideal voltmeter (V) shown in Fig.

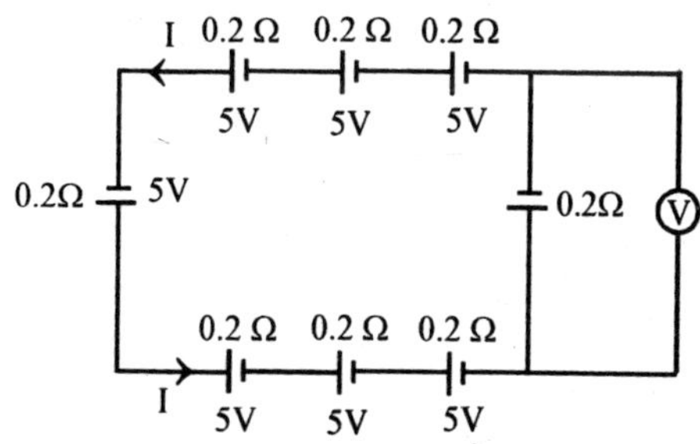

Solution:

Current through each cell,

$$I = \frac{5 \times 8}{0.2 \times 8} = \frac{40}{1.6} = 25\text{A}$$

Reading of voltmeter = terminal voltage across a cell (V)

$$= E - I\,r = 5 - 25 \times 0.2 = 0\ \text{V}.$$

Example 38:

ABCD is a uniform circular wire of resistance 2 Ω. AOC and BOD are two wires (each of resistance 1 Ω) forming diameters at rt. angles to each other. Show that the resistance of the net work is 15/14 Ω, if the battery is placed in AD.

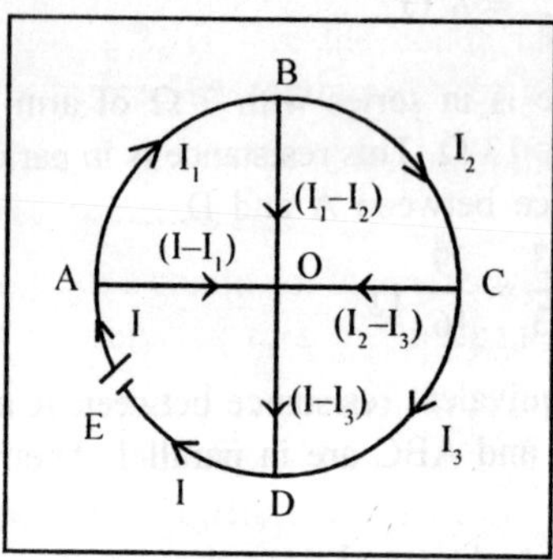

Solution:

Let E be the emf of the battery used. Let the distribution of current in the various arms be as shown in circuit.

Here, $R_{AB} = R_{BC} = R_{CD}$

$$= R_{DA} = R_{AO} = R_{OD}$$

$$= R_{OC} = R_{OD} = \frac{1}{2}\Omega$$

Applying Kirchhoff's second law, we have.

In closed loop ABOA,

$$I_1r + (I_1 - I_2)r - (I - I_1)r = 0$$

or $$3I_1 - I_2 = I \quad \text{...(i)}$$

In closed loop BCOB,

$$I_2r + (I_2 - I_3)r - (I_1 - I_2)r = 0$$

or $$-I_1 + 3I_2 - I_3 = 0 \quad \text{...(ii)}$$

In closed loop CDOC,

$$I_3r - (I - I_3)r - (I_2 - I_3)r = 0$$

or $$-I_2 + 3I_3 = I$$

or $$I_2 = 3I_3 - I \quad \text{...(iii)}$$

Putting this value in (i), we get

$$3I_1 - (3I_3 - I) = I$$

or $$3I_1 - 3I_3 = 0$$

or $$I_1 = I_3$$

Putting value of I_2 and I_3 in (ii), we get

$$-I_1 + 3(3I_1 - I) - I_1 = 0$$

or $$7I_1 = 3I$$

or $$I_1 = \frac{3}{7}I$$

In closed loop AODA,

$E = r(I - I_1) + r(I - I_3) - I_1 = 0$

or $$E = \frac{1}{2}\left(1 - \frac{3}{7}I\right) + \frac{1}{2}\left(1 - \frac{3}{7}I\right) + \frac{1}{2}I$$

$$= \frac{15}{14}I \quad \text{...(iv)}$$

If R is the effective resistance of net work, then E = IR ...(v)

From (iv) and (v), $$IR = \frac{15}{14}I$$

or $$R = \frac{15}{14}\Omega$$

Example 39:

In Fig., illustrates a potentiometer circuit by means of which we can vary a voltage V applied to a certain device possessing a resistance R. The potentiometer has a length l and a resistance R_0 and voltage V_0 is applied to its terminals. Find the voltage V fed to the device as a function of distance x. Analyses separately the case $R >> R_0$.

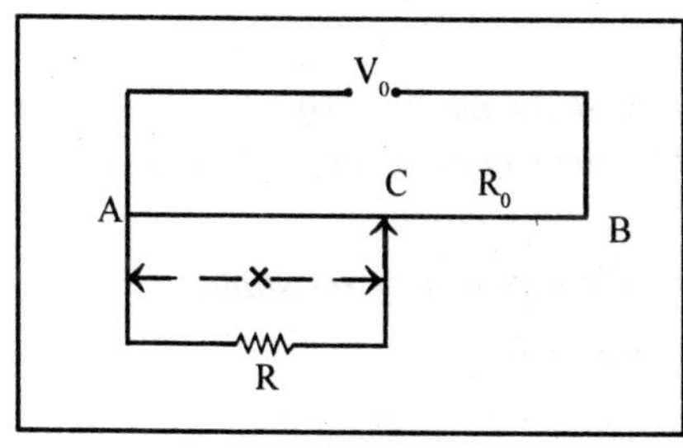

Solution:

Resistance per unit length of potentiometer wire = R_0/l.

Total resistance between A and B

= resistance between B and C + resistance between A and C

$$= \frac{(l-x)\,R_0}{l} + \frac{R \times \frac{R_0}{l}\,x}{R \times \frac{R_0}{l}\,x}$$

$$= \left(\frac{(l-x)}{l}\right) R_0 + \frac{xRR_0}{lR + R_0x}$$

$$= R_0\left[\left(\frac{(l-x)}{l}\right) + \frac{xR}{lR + R_0x}\right]$$

Voltage V fed to the device = voltage drop per unit resistance × resistance between A and C.

$$\therefore\ V = \frac{V_0 \times xR\,R_0 / (lR + R_0x)}{R_0\left[\frac{l-x}{l} + \frac{xR}{lR + R_0x}\right]}$$

$$= \frac{V_0\,R\,x}{(lR + R_0x)\left[1 - \frac{x}{l} + \frac{xR}{lR + R_0x}\right]}$$

$$= \frac{V_0\,R\,x}{(lR + R_0x)\left(1 - \frac{x}{l}\right)}$$

For $R >> R_0$, $V = \frac{V_0\,R\,x}{lR} = \frac{V_0x}{l}$.

Example 40:

A metal wire of diameter 2×10^{-3} m and length 100 m has a resistance of 0.5475 ohms at 20° C and 0.805 ohms at 150° C. Find the values of

(i) temperature coefficient of resistance

(ii) its resistance at 0° C

(iii) its resistivities at 0° C and 20° C

Solution :

If R_{20} and R_{150} are the resistances of the wire at 20° C and 150° C respectively and α is the temperature coefficient of resistance of the metal, then

$R_{20} = R_0 (1 + \alpha \times 20) = 0.5475$ ...(1)

$R_{150} = R_0 (1 + \alpha \times 150) = 0.8050$ ...(2)

Dividing equation (1) by equation (2)

$$\frac{1+20\alpha}{1+150\alpha} = \frac{0.5475}{0.8050}$$

$$\therefore \quad \alpha = \frac{0.8050 - 0.5475}{0.5475\times150 - 0.8050\times20} = \mathbf{3.9\times10^{-3}{}^{\circ}C}$$

(ii) Putting value of α in equation (1), we get

$$R_0 = \frac{0.5475}{1+\alpha\times20} = \frac{0.5475}{1+3.9\times10^{-3}\times20}$$

= **0.5079 ohms.**

(iii) Now $R_0 = \rho_0 (l / A)$

$$\therefore \; \rho_0 = \frac{R_0A}{l} = \frac{0.5079\times\pi\times10^{-6}}{100}$$

$= 1.596 \times 10^{-8}$ ohms × meter

and $\rho_{20} = \rho_0 (1 + \alpha \times 20) = 1.596 \times 10^{-8}[1 + 3.9 \times 10^{-3} \times 20]$

$= \mathbf{1.720 \times 10^{-8}}$ **ohms × meter**

Example 41:

Calculate the following for a 100 W – 250 V bulb.

(i) Resistance of the filament of the bulb.

(ii) Current through bulb when it is connected to 250 volt line.

(iii) Power consumed when bulb is connected to a 200 volt line.

Solution :

(i) $R = \frac{V^2}{P} = \frac{250\times250}{100}$ = **625 ohms**

(ii) Current through bulb

$$i = \frac{P}{V} = \frac{100 \text{ watt}}{250 \text{ Volt}} = \mathbf{0.4\ amp.}$$

(iii) Power consumed when connected to a 200 volt line

$$P' = \frac{V^2}{R} = \frac{200 \times 200}{625} = \mathbf{64\ Watt}$$

Example 42:

One Kilowatt electric heater is to be used with 220 volt DC supply.

(i) What is the current in the heater ?

(ii) What is its resistance ?

(iii) What is the power dissipated in the heater ?

(iv) How much heat in calories is produced/sec?

(v) How many grams of water at 100 ° C will be converted per minute into steam at 100 ° C with the heater ? Assume that the heat losses due to radiaion are negligible. Latent heat of steam = 540 cal/gram.

Solution :

P = 1000 Watt, V = 220 volts.

(i) Current in the heater i = (P/V) = (1000 W/220 V) = 4.55 A

(ii) Resistance of the heater element

$$R = \frac{V^2}{P} = \frac{220 \times 220}{1000} = \mathbf{48.4\ ohms}$$

(iii) Power consumed by the heater = 1000 Watt

(iv) Heat produced per sec.

$$H = \frac{P \times t}{J} = \frac{1000 \times 1}{4.2} = \mathbf{240\ cal/sec}$$

(v) Heat produced in one minute

= 240 × 60 ; Q = 14400 cal

Hence Q = ML

i.e. Mass of water at 100° C converted into steam at 100° C

= M = Q/L= (14400 / 540)

= 26.8 gms.

Example 43:

Two Bulbs rated at 60 W – 240 V and 100 W – 200 V, are connected in series across a 240 volt line. If the resistance of two bulbs remain constant, what will be the potential difference across each bulb ?

Solution :

Resistance of 1st bulb $R_1 = \frac{V^2}{P_1} = \frac{240 \times 240}{60}$ = **960 ohms**

Resistance of 2nd bulb $R_2 = \frac{V^2}{P_2} = \frac{200 \times 200}{100}$ = **400 ohms**

As two bulbs are connected in series, hence their total resistance

$R = R_1 + R_2 = 960 + 400 =$ **1360 ohms.**

Note : We will study in next chapter, that effective resistance, when two or more resistance are connected in series, is given by : $R = R_1 + R_2 + ...$

Hence current flowing through each resistance (current remains same in series combination of resistances)

$i = \frac{V}{R_1 + R_2} = \frac{240}{1360}$ amp

$\therefore V_1$ = potential difference across 1st bulb

$= iR_1 = \frac{240}{1360} \times 960$ = **169.4 Volts**

and V_2 = potential difference across 2nd bulb

$= iR_2 = \frac{240}{1360} \times 400$ = **70.6 Volts**

Example 44:

Two tungston lamps with resistances R_1, R_2 respectively at full incandescence are connected, first in parallel and then in series, in a lighting circuit of negligible internal resistance. If $R_1 > R_2$,

(i) which lamp will glow more brightly when they are connected in parallel ?

(ii) If the lamp of resistance R_1 now burns out, how will the illumination produced change ?

(iii) Which lamp will glow more brightly when they are connected in series ?

(iv) If the lamp of resistance R_2 now burns out, and the lamp of resistance R_1 alone is plugged in, will the illumination increase or decrease?

(v) In any case, would physically bending a supply wire cause any change in the illumination?

Solution :

(i) when the lamps are connected in parallel, then potential difference V across each lamp will be same and will be equal to potential difference necessary for full brightness of each bulb.

Because illumination produced by a lamp is proportional to electric power consumed in it, and power consumed

$$P_1 = \frac{V^2}{R_1} < \frac{V^2}{R_2} = P_2$$

hence *illumination produced by 2nd bulb will be higher than produced by 1st bulb i.e. bulb having lower resistance will shine more brightly.*

(ii) When R_1 burns out, then power is dissipated in R_2 only. Because internal resistance is quite low in lighting circuit, potential difference is still equal to V, hence power dissipated in 2nd lamp i.e.

$$\frac{V^2}{R_2} < \left(\frac{V^2}{R_1} + \frac{V^2}{R_2}\right) \quad \text{i.e. net power consumed initially.}$$

In other words, *net illumination will now decrease.*

(iii) When the two lamps are connected in series, the potential difference across each lamp will be different but current i flowing through each lamp will be same.

Hence $P_1 = i^2 R_1 > i^2 R_2 = P_2$

i.e. illumination produced by 1st lamp will be higher as compared to that produced by 2nd lamp i.e. *lamp having higher resistance will glow more brightly.*

(iv) When lamp of resistance R_2 burns out and only lamp of resistance R_1 is connected in the circuit, then current flowing through the circuit will change. Let new current is i.

Because potential difference still remains same (due to low internal resistance), hence

$i' R_1 = i(R_1 + R_2)$

or $i' = [i (R_1 + R_2)]/R_1$

If P' is the power consumed, then

$P' = i'^2 R_1 = [i^2 (R_1 + R_2) . (R_1 + R_2)]/R_1$

When both the lamps were present then total power consumed was given by

$P_S = P_1 + P_2 = i^2 (R_1 + R_2)$ i.e. $P' > P_S$

i.e. *Illumination gets increased when only one bulb is used.*

(v) If a water pipe is given a bend at some point, then it definitely reduces the flow of water in the pipe but this is not true in case of an electric current flowing in a conductor because electric current is established in a conductor due to drift motion of electron in it along the line of the potential gradient. Hence illumination is not effected due to bending along the length of supply wires.

Example 45:

An electric kettle rated at (1500 W - 240 V) takes 7 minutes to increase the temperature of 1.8 kg of water from 20° C to100 ° C.

(a) determine the efficiency of the kettle and

(b) find out the mass of water which may be boiled by 1KWH of electrical energy.

The specific heat capacity of water is 4.18×10^3 Joules/kg/° K

Solution :

Energy supplied to kettle i.e. input Energy

$= P \times t = 1500 \times 7 \times 60 = 630 \times 10^3$ J

Energy obtained from kettle i.e. output energy = ms.dθ

$= 1.8 \times 4.18 \times 10^3 \times (100 - 20) = 602 \times 10^3$J

(a) Hence efficiency of kettle

$$\eta = \frac{\text{Output Energy}}{\text{Input Energy}} \times 100\% = \frac{602}{630} \times 100 = \mathbf{95.5\%}$$

(b) As Energy required to boil 1.8 kg of water is 630×10^3 J. Hence mass of water boiled by 1 KWH of energy

$$= \frac{1.8 \times 3.6 \times 10^6}{630 \times 10^3} \text{ Kg} = \mathbf{10.29\ Kg.}$$

Example 46:

A generator of e.m.f. 80 volt has an internal resistance of 0.04 ohms. If the terminal voltage is 75 volts, calculate

(i) the current

(ii) power supplied

(iii) power dissipated in the generator

(iv) reading of an ammeter of resistance 1 ohm connected in the circuit.

Solution :

(i) $i = \frac{E - V}{r} = \frac{80 - 75}{.04} = 125$ amp.

(ii) Power suppled P = Vi = 125 × 75 = **9375 Watt**

(iii) Power dissipated P' = i^2 r = $(125)^2 \times 0.04$ = **625 Watt**

(iv) Resistance $R = \frac{V}{i} = \frac{75}{125}$ = **0.6 ohms**

Total external resistance of the circuit R'= 0.6 | 1 = 1.6 ohms

∴ current in the circuit or Reading of ammeter

$$i' = \frac{E}{R' + r} = \frac{80}{1.6 + 0.04} = \frac{80}{1.64} = \mathbf{48.78\ Amp}$$

Example 47(a):

A fuse of lead wire has an area of cross section 0.2 mm². On short circuiting, the current in the fuse wire reaches 30 amp. How long after the short circuiting, will fuse begin to melt ? For lead, specific heat 0.032 cal / gm/° C, melting point = 327 ° C, density = 11.34 gm/cm³ and resistivity = 22 × 10⁻⁶ ohms-cm. The initial temperature of wire is 20° C. Neglect heat losses.

Solution :

Given that, Resistivity $\rho = 22 \times 10^{-6}$ ohms-cm = 22×10^{-8} ohms-m;

$A = 0.2\ mm^2 = 0.2 \times 10^{-6}\ m^2 = 2 \times 10^{-7}\ m^2$;

$(\theta_2 - \theta_1) = 327 - 20 = 307°\ C$;

$d = 11.34\ gms/cm^3 = 11.34 \times 10^3\ kg/m^3$ current $i = 30$ amp, specific heat $s = 0.032$ cal/gm $-$ °C $= 0.032$ K cal/kg- ° C.

For the melting of fuse wire,

$$\left[\begin{array}{l}\text{Heat produced due to} \\ \text{flow of current in fuse} \\ \text{wire in time t.}\end{array}\right] = \left[\begin{array}{l}\text{Heat required for increasing} \\ \text{the temperature of fuse wire} \\ \text{upto melting point.}\end{array}\right]$$

$$\frac{i^2Rt}{4.2} = Q = ms\,(\theta_2 - \theta_1)$$

or $$\frac{i^2\,(\rho l/A)\,t}{4.2 \times 10^3} = (A/d)\ s\ (\theta_2 - \theta_1)$$

$$\left[\frac{(\text{amp})^2 \times [\text{ohm} \times \text{meter} \times (\text{meter/meter}^2)]\text{sec}}{4.2 \times 10^3}\right.$$

$$\left. = \text{meter}^2 \times \text{meter} \times \frac{\text{kg}}{\text{meter}^3} \times \frac{\text{K cal}}{\text{kg} - °\text{C}} \times °\text{C}\right]$$

[K cal = K cal]

Thus, $$\frac{i^2\rho t}{4.2 \times 10^3\ A} = A\ d\ s\ (\theta_2 - \theta_1)$$

$$\therefore\ t = \frac{4.2 \times 10^3\ A^2\ ds(\theta_2 - \theta_1)}{i^2\rho}$$

$$= \frac{4.2 \times 10^3 \times (2 \times 10^{-7})^2 \times (11.34 \times 10^3) \times (0.032) \times 307}{(30)^2 \times 22 \times 10^{-8}}$$

= **0.095 seconds**

Example 47(b):

Forty electric bulbs are connected in series across a 220 V supply. After one bulb is fused, the remaining 39 are connected again in series across the same supply. In which case will there be more illumination and why ?

Solution :

If R is the resistance of each bulb, then total resistance of 40 bulbs in series will be 40R ohms. When connected across a supply voltage V, the total power consumed by the system or illumination produced is given by $P_{40} = V^2 / 40R$

When one of the bulbs is fused, the resistance of the remaining 39 bulbs in series = 39R and power consumed by the system in now given by $P_{39} = V^2/39R$

Thus it is clear that $\dfrac{V^2}{30R} > \dfrac{V^2}{40R}$

i.e. illumination produced by 39 bulbs will be higher.

Example 48:

Two uniform wires of the same material, each weighing 1 gm but one having double the length of the other, are connected in series. They carry a current of 10 A. The length of longer wire is 20cm. Calculate the rate of consumption of energy in each of the two wires. Which one will get hotter ? Density of the material of the wire = 11.34 gm/cm³ . Specific resistance of the material = 20.6 × 10⁻⁵ ohms × cm.

Solution :

Because mass and density of each wire are same, hence volume of each wire

$V = (1.0/11.34)\ cm^3$

If A represents the area of cross section of the longer wire of length 20cm, then

$$20\ A = \left[\frac{1.0}{11.34}\right]$$

$$\text{or} \quad A = \frac{1}{20 \times 11.34}\ cm^2$$

Hence area of cross-section of the shorter wire = 2A

(C.S. area is doubled because length of shorter wire is half the length of longer wire). Now Resistance of longer wire.

$$R_1 = \rho.\frac{l}{A} = \frac{(20.6 \times 10^{-6}) \times 20}{I} \times (20 \times 11.34)$$

= 0.09344 ohms

∴ Power consumed by the longer wire

$P_1 = i^2R_1 = (10)^2 \times (.09344) =$ **9.344 Watt**.

Resistance of the shorter wire

$$R_2 = \rho\frac{(l/2)}{2A} = \frac{1}{4}\frac{(\rho l)}{A} = \frac{R_1}{4}$$

∴ Power consumed by the shorter wire

$$P_2 = i^2R_2 = i^2\frac{R_1}{4} = \frac{P_1}{4} = \frac{9.344}{4} = \mathbf{2.34\ Watt}$$

i.e. $P_2 < P_1$ i.e. **longer wire becomes hotter.**

Example 49(a):

A standard 50 W electric lamp in series with a room heater is connected across the mains. If the 50 W bulb is replaced by a 100 W bulb, will the heater output be larger, smaller or remain the same?

Solution :

We know that Resistance of 50 W bulb > Resistance of 100 W bulb ($\because R \propto 1/P$), hence when 50 W bulb is replaced by 100 W bulb in the circuit, the total resistance of the circuit is decreased and a higher current now flows in the circuit i.e. power consumed (= i^2 R) by heater is increased or **heater output is increased.**

Example 49(b):

What current must be passed through a 40Ω immersion heater, dipped in a tank of heat capacity 630 J/K, containing 2.75 kg of water, so that heat developed increases the temperature of the tank from 10°C to 40°C in five minutes. Assume that 4200 Joules of work is equivalant to 1k cal.

Solution :

Because heat capacity of tank or ms = 630 Joules/ °K = 630 Joules/ °C = (630/4200) kcal/°C, = 0.15 k cal/°C, hence quantity of heat required to raise its temperature by 1°C = 0.15 k cal which is just equal to the quantity of heat required to raise the temperature of 0.15 kg of water by the same amount i.e.1°C.

Hence, effective mass of water = 2.75 + 0.15 = 2.9 kg.

and heat required to raise its temperature by 30°C = Q = ms.dθ
= 2.9 kg × (1 k cal/kg – °C) × 30°C
= 2.9 × 30 × k cal = 2.9 × 30 × 4200 Joules

If i is the current which passes through immersion heater, then i^2 Rt = Q

$$\therefore \; i^2 = \frac{Q}{tR} = \frac{2.9\times30\times4200}{5\times60\times40} = 30.45$$

Hence i = **55.18 amp**

Example 50(a):

A two meter long wire of resistance 4 ohms and diameter 0.64 mm is coated with plastic insulation of thickness 0.06 mm. When a current of 5 amp is passed through the wire, what will be temperature difference across the insulation in steady state. Coefficient of thermal conductivity of the insulating material = 0.16 × 10^{-2} cal/sec/cm/° C and 1 cal = 4.2J

Solution :

Given that Radius of the metal wire = R = 0.32 mm = .032 cm, Thickness of insulating material = d = .06 mm = .006 cm, Inner radius of insulation = R = .032 cm, Outer radius of insulation = R + d = 0.032 + .006 = .038 cm, Average radius of insulation = r = $\left(\frac{1}{2}\right)$ (.038 + .032) = .035 cm

Curved surface area, corresponding to average radius = A = 2πrL (where L is the length of wire) Heat energy that passes radially through insulating material, may be assumed as passing through curved surface area only because base area is negligible as compared to curved surface area.

In steady state :

$$\left.\begin{matrix}\text{Heat produced per sec in the metallic}\\ \text{wire due to flow of current}\end{matrix}\right] = \left[\begin{matrix}\text{Heat conducted per second radially}\\ \text{through insulating material}\end{matrix}\right.$$

$$\frac{i^2R}{4.2}\text{ cal} = \frac{KA(\theta_2-\theta_1)}{a}$$

$$\therefore (\theta_2 - \theta_1) = \frac{i^2Rd}{4.2KA} = \frac{i^2Rd}{4.2K \times 2\pi rL}$$

$$= \frac{(5)^2 \times 4 \times .006}{4.2 \times (0.16 \times 10^{-2}) \times (2\pi \times .035 \times 200)}$$

$$= \frac{23.8 \times .006}{0.16 \times 10^{-2} \times 44} = \mathbf{2^\circ C}$$

Example 50(b):

Two bulbs rated 25 W - 220 V and 100 W - 220 V are connected in series to a 440 volt supply. Show with necessary calculations, which bulb, if any, will fuse ? What will happen if the two bulbs are connected in parallel to the same supply ?

Solution :

Given that P_1 = 25 Watt, P_2 = 100 Watt and voltage V = 220 volt.

Hence, resistances of the filaments of two bulbs are given by

$R_1 = (V^2/P_1) = (220 \times 220/25) = 1936$ ohms

& $R_2 = (V^2/P_2) = (220 \times 220/100) = 484$ ohms

Now, maximum currents that can be allowed to pass through the two bulbs, are given by

i_1 (max) = $(P_1/V) = (25/220) = 0.114$ amp

i_2 (max) = $(P_2/V) = (100/220) = 0.454$ amp

Case I : *When bulbs are connected in series,* then total resistance $R_5 = R_1 + R_2 = 2420$ ohms, and current in the circuit = current through each bulb = $i_S = (V'/R_S) = (440/2420) = 0.18$ **amp.**

As, $i_S > i_1$ (max) but $< i_2$ (max), hence *bulb rated 25 W – 220 V, will fuse.*

Case II : When bulbs are connected in parallel, then potential difference across each bulb will be same and equal to 440 volt.

Hence current through 25 Watt bulb

$i_1 = \frac{V'}{R_1} = \frac{440}{1936}$ = 0.23 amp & that through 100 Watt bulb

$$i_2 = \frac{V'}{R_2} = \frac{440}{484} = 0.91 \text{ amp}$$

As $i_1 > i_1$ (max) and i_2 (max), hence *both bulbs will fuse.*

Example 51:

It is required to send a current of 10 amp through a resistance $R = 3\Omega$.

(a) What is the minimum number of cells required if each cell has an e.m.f. of 10 V and internal resistance 1 ohm.

(b) Find the power dissipated in R.

Solution :

(a) Suppose N be the required number. Let us arrange in m rows each containing n cells in series. Then mn = N.

The internal resistance of the battery formed

$= \frac{nr}{m} = \frac{n}{m}$ and e.m.f. of the battery = nE = 10n

Then, $$10 = \frac{10n}{\frac{n}{m}+3}$$

or, $10 = \dfrac{10\,mn}{n+3m}$ or, n + 3m = mn

or, $n+3\dfrac{N}{n}$ or, $n^2 + 3\,N = Nn$

Differentiating with respect to n,

$$2n+3\frac{dN}{dn} = \frac{dN}{dn} \times n + N$$

When N is minimum $\dfrac{dN}{dn} = 0$,

Hence $n = \dfrac{N}{2}$

Hence, when N is minimum, $\dfrac{N^2}{4} + 3N = \dfrac{N^2}{2}$

or, $\dfrac{N^2}{4} = 3N$ or, $\mathbf{N = 12}$

Example 52:

Two coils when joined in series have an effective resistance of 18 ohms but when joined in parallel, an effective resistance of 4 ohms. Calculate the values of their resistances.

Solution :

If R_1 and R_2 are the resistances of two coils, then in series combination their effective resistance

$R_S = R_1 + R_2$ i.e. $R_1 + R_2 = 18$ ...(1)

In parallel combination, their effective resistance

$$R_P = \frac{R_1R_2}{R_1+R_2} \text{ i.e. } \frac{R_1R_2}{R_1+R_2} = 4$$

or $R_1R_2 = 72$ ...(2)

From eq. (1) and (2)

$R_1 - R_2 = \sqrt{[(R_1 + R_2)^2 - 4R_1R_2]}$

$= \sqrt{[(18)^2 - 4 \times 72]} = 6$ ...(3)

From eq. (1) and (3) :-

R_1 = 12 ohms & R_2 = 6 ohms.

Multiple Choice Questions

1. A current of 2 ampere flowing through a conductor produces 80 joule of heat in ten seconds. The resistance of the conductor is

   (a) 0.5 Ω (b) 2 Ω

   (c) 4 Ω (d) 20 Ω

2. Two identical batteries, each of e.m.f. 2 volt and internal resistance 1 ohm are available to produce heat in a resistance R = 0.5 ohm, by passing a current through it. The maximum joulean power that can be developed across R using these batteries is

   (a) (8/9) W (b) 1.28 W

   (c) 2.0 W (d) 3.2 W

3. A galvanometer of resistance 10 W gives full-scale deflection when 1 mA current passes through it. The resistance required to convert it into a voltmeter reading upto 2.5 V is

(a) 24.9 Ω (b) 249 Ω

(c) 2490 Ω (d) 24900 Ω

4. Neutral temperature for a thermocouple is 270°C. Temperature of cold junction is 150°C. What is the temperature of inversion?

(a) 255°C (b) 285°C

(c) 575°C (d) 525°C

5. A 5°C rise in temp. is observed in a conductor by passing a current. When the current is doubled the rise in temperature will be approximately

(a) 10°C (b) 12°C

(c) 16°C (d) 20°C

6. A new flashlight cell with an emf of 1.5 V gives a current of 15 A when connected directly to an ammeter of resistance 0.04 Ω. The internal resistance of the cell in ohms is

(a) 0.04 (b) 0.06

(c) 0.10 (d) 10

7. A circuit with two cells in opposition to each other. One cell has an emf of 6 V and internal resistance of 2 W and the other cell has an emf of 4 V and internal resistance of 8 Ω. The potential difference across the terminals X and Y is

(a) 5.4 V (b) 5.6 V

(c) 5.8 V (d) 6.0 V

8. An electric bulb is rated 220 V and 100 watt. Power consumed by it when operated on 110 volt is

(a) 25 watt (b) 50 watt

(c) 75 watt (d) 90 watt

9. A constant voltage is applied between two ends of a uniform conducting wire. If both the length and radius of the wire are doubled

(a) the heat produced in the wire will be doubled

(b) the electric field across the wire will be doubled

(c) the heat produced will remain unchanged

(d) the electric field across the wire will become one half

10. A heat of 220 V heats a volume of water in 5 minutes time. A heater of 110 V heats the same volume of water in

    (a) 5 minutes (b) 8 minutes
    (c) 10 minutes (d) 20 minutes

11. If the current in the electric bulb decreases by 0.5 percent, then the power in the bulb decreases by approximately

    (a) 0.25% (b) 0.5%
    (c) 1% (d) 2%

12. If two electric bulbs whose resistance are in the ratio of 1 : 2 are connected in parallel to a constant voltage source. The powers dissipated in them have the ratio

    (a) 2 : 1 (b) 1 : 2
    (c) 1 : 1 (d) 1 : 4

13. If R_1 and R_2 are the filament resistances of 200 watt and a 100 watt bulb respectively, both designed to run at the same voltage, then

    (a) R_1 is four times R_2
    (b) R_2 is four times R_1
    (c) R_1 is two times $3R_2$
    (d) R_2 is two times R_1

14. If two bulbs of wattage 25 Ω and 100 Ω respectively each rated at 220 V connected in series with the supply of 440 V, which bulb will fuse ?

    (a) 25 W bulb (b) 100 W bulb
    (c) both of them (d) none of them

15. Amount of charge in coulomb required to deposit one gram equivalent of substance by electrolysis is

    (a) 9.6×10^4 (b) 4.8×10^{-4}
    (c) 96500 (d) 6500

16. N identical cells, each of emf E, are connected in parallel. The emf of the combination is

    (a) NE (b) E
    (c) N^2 E (d) E/N

17. Two heater wires of equal length are first connected in series and then in parallel. The ratio of heat produced in the two cases is
    (a) 2 : 1 (b) 1 : 2
    (c) 4 : 1 (d) 1 : 4
18. A short-circuited coil is placed in a time-varying magnetic field. Electrical power is dissipated due to the current induced in the coil. If the number of turns were to be quadrupled and the wire radius halved, the electrical power dissipated would be
    (a) halved (b) the same
    (c) doubled (d) quadrupled
19. Two wires of the same material and having same uniform area of cross-section are connected in an electrical circuit. The masses of the wires are m and 2 m respectively. When a current I flows through the circuit, the heat produced in them in a given time is in the ratio
    (a) 2 : 1 when they are connected in series
    (b) 2 : 1 when they are connected in parallel
    (c) 1 : 2 when they are connected in parallel
    (d) 1 : 2 when they are connected in series
20. The terminal voltage of a battery is
    (a) always equal to its emf
    (b) always less than its emf
    (c) greater or less than its emf depending on the direction of the current through the battery
    (d) greater of less than its emf depending on the magnitude of its internal resistance
21. The e.m.f. of the battery in a thermocouple is doubled. The rate of heat generation at one of the junctions will
    (a) become half (b) remain unchanged
    (c) become double (d) become four times
22. A constant voltage is applied between the two ends of a uniform metallic wire. Some heat is developed in it. The heat developed is doubled if

(a) the radius of the wire is doubled
(b) the length of the wire is doubled
(c) both the length and radius of the wire are halved
(d) both the length and radius of the wire are doubled

23. In an electroplating experiment, m gram of silver is deposited when 4 ampere of current flows for 2 minutes. The amount (in gram) of silver deposited by 6 ampere of current for 40 seconds will be

(a) (m/4) (b) (m/2)
(c) 2m (d) 4m

24. Metals are good conductor of heat than insulator because

(a) their atoms are relatively apart
(b) they contain free electrons
(c) their atom collide frequency
(d) they have reflecting surface

25. Ohm's law does not hold good for

(a) current through an N-P-N transistor
(b) current through a column of mercury
(c) current flowing in a diode valve
(d) any of the above

26. A micrometer has resistance of 100 ohm and full-scale range of 50 μA. It can be used as a voltmeter or as higher range ammeter, provided a resistance is added to it. Pick the correct range and resistance combination(s)

(a) 50 V range with 10 k ohm resistance in series
(b) 10 V range with 200 k ohm resistance in series
(c) 5 mA range with 1 ohm resistance in parallel
(d) 10 mA range with 1 ohm resistance in parallel

27. Two electric lamps of 40 watt each are connected in parallel. The power consumed by the combination will be

(a) 20 watt (b) 60 watt
(c) 80 watt (d) 100 watt

28. A copper voltameter is connected in series with a heating coil of resistance 10 Ω. A steady current flows in the circuit for 20 minutes and deposits 0.99 g of copper. If the E.C.E. of copper is 0.00033 g C^{-1}, the amount of heat generated in the coil in this time is

(a) 2.5×10^4 J (b) 5.0×10^4 J

(c) 7.5×10^4 J (d) 10.0×10^4 J

29. When current is passed through a junction of two dissimilar metals, heat is evolved or absorbed at the junction. This process is called

(a) Seebeck effect (b) Joule effect

(c) Peltier effect (d) Thomson's effect

30. If 2.2 kilo watt power is transmitted through a 10 Ω line at 22000 volt, the power loss in the form of heat will be

(a) 0.1 watt (b) 1 watt

(c) 10 watt (d) 100 watt

31. 1°C rise in temperature is observed in a conductor by passing a certain current. If the current. If the current is doubled then the rise in temperature is approximately

(a) 1°C (b) 2°C

(c) 2.5°C (d) 4°C

32. If a power of 100 W is being supplied across a potential difference of 200 V, current flowing is

(a) 0.5 A (b) 1 A

(c) 2 A (d) 20 A

33. A current of 1.0 A is passed through a coil across which is a potential difference of 210 V. The coil which is embedded in ice will melt .... of ice per hour

(a) 0.84 kg (b) 8.4 kg

(c) 84 kg (d) none of these

34. A condenser having a capacity 2.0 μF is charged to 200 V and then the plates of the capacitor are connected to a resistance wire. The heat produced in joules will be

(a) 2×10^{-2} J (b) 4×10^{-2} J

(c) 2×10^{4} J (d) 2×10^{10} J

35. How many coulombs of electricity must pass through acidulated water to liberate 22.4 litres of hydrogen at N.T.P. ?

(a) 1 C (b) 1.6×10^{-19} C

(c) 96500 C (d) 193000 C

36. A nichrome wire 50 cm long and 1 mm^2 in cross-section carries a current of 4 A when connected to a 2 V storage battery. The resistivity of nichrome is

(a) 1×10^{-6} Ωm (b) 2×10^{-7} Ωm

(c) 4×10^{-7} Ωm (d) 5×10^{-7} Ωm

37. The current density vector due to flow of electrons

(a) + nev (b) – nev

(c) $\frac{nv}{e}$ (d) $\frac{ev}{n}$

38. A DC milliammeter has resistance of 12 Ω and gives a full scale deflection for a current of 0.01 A. To convert it into a voltmeter giving a full-scale deflection for 3 V, the resistance required to be put in series with the instrument is

(a) 102 Ω (b) 288 Ω

(c) 300 Ω (d) 412 Ω

39. Consider the following two statements

(i) free-electron density is different in different metals.

(ii) free-electron density in a metal depends on temperature.

Seebeck effect is caused

(a) due to (i) but not due to (ii)

(b) due to (ii) but not due to (i)

(c) due to both (i) and (ii)

(d) neither due to (i) nor due to (ii)

40. Considering the statements (i) and (ii) in question 246, Peltier effect is caused

(a) due to (i) but not due to (ii)

(b) due to (ii) but not due to (i)

(c) due to both (i) and (ii)

(d) neither due to (i) nor done to (ii)

41. Considering the statements (i) and (ii) in question 246, Thomson effect is caused

(a) due to (i) but not due to (ii)

(b) due to (ii) but not due to (i)

(c) due to both (i) and (ii)

(d) neither due to (i) nor done to (ii)

42. The neutral temperature of copper iron thermocouple is 270°C. If the temperature of cold junction is 30°C, the temperature of inversion will be

(a) 510 K (b) 510°C

(c) 300 K (d) 300°C

43. An electric bulb is rated 220 V, 60 W. Its electrical resistance is nearly

(a) 0.807 Ω (b) 4 Ω

(c) 708 Ω (d) 807 Ω

44. A capacitor with no dielectric is connected to a battery t = 0. Consider a point A in the connecting wires and a point B in between the plates

(a) there is no current through A

(b) there is no current through B

(c) there is a current through A as long A as long as charging is not complete

(d) there is current through B as long as the charging is not complete

45. Twelve equal wires, each of resistance 4Ω are jointed to form at skeleton cube. The effective resistance of the cube between two points on the same edge of the cube is

(a) 7/3 Ω (b) 28/5 Ω

(c) 10/3 Ω (d) 5/3 Ω

46. In an experiment to measure the internal resistance of a cell by potentiometer, it is found that the balance point is at a length of

2 metre when the cell is shunted by a 5 Ω resistor and is at a length of 3 metre when the cell is shunted by a resister 10 ohm. The internal resistance of the cell is then

(a) 1.5 ohm (b) 10 ohm

(c) 15 ohm (d) 1 ohm

47. A capacitor with no dielectric is connected to a battery at $t = 0$. Consider a point A in the connecting wires and a point B in between the plates

(a) there is no current through A

(b) there is no current through B

(c) there is a current through A as long as the charging is not complete

(d) there is a current through B as long as the charging is not complete

48. A wire has resistance of 10 Ω. It is stretched by one-tenth of its original length. Then its resistance will be

(a) 9 Ω (b) 10 Ω

(c) 11 Ω (d) 12.1 Ω

49. Consider the following two statements :

(A) Kirchoff's junction law follows from conservation of charge

(B) Kirchoff's loop law follows from conservative nature of electric field

(a) both A and B are correct

(b) A is correct but B is wrong

(c) B is correct but A is wrong

(d) both A and B are wrong

50. Consider a capacitor-charging circuit. Let Q_1 be the charge given to the capacitor in a time interval of 10 ms and Q_2 be the charge given in the next time interval of 10 ms. Let 10 μC charge be deposited in a time interval t_1 and the next 10 μC charge is deposited in the next time interval t_2

(a) $Q_1 > Q_2$, $t_1 > t_2$ (b) $Q_1 > Q_2$, $t_1 < t_2$

(c) $Q_1 < Q_2$, $t_1 > t_2$ (d) $Q_1 < Q_2$, $t_1 < t_2$

51. A certain piece of copper is to be shaped into a conductor of minimum resistance. Its length and diameter should be

(a) L, D (b) 2L, D

(c) L/2, 2 D (d) 2L, D/L

52. Two resistors R and 2R are connected in series in an electric circuit. The thermal energy developed in R and 2 R are in the ratio

(a) 1 : 2 (b) 2 : 1

(c) 1 : 4 (d) 4 : 1

53. Which of the following quantities do not change when a resistor connected to a battery is heated due to the current ?

(a) drift speed

(b) resistivity

(c) resistance

(d) number of free electrons

54. When a potential difference is applied across the ends of a linear conductor, the free electrons

(a) are accelerated continuously from the lower potential end of the higher potential end of the conductor

(b) are accelerated continuously from the higher potential end of the lower potential end of the conductor

(c) acquire a constant drift velocity from the lower potential end to the higher potential end of the conductor

(d) are set in motion from their position of rest

55. The net resistance of a voltmeter should be large to ensure that

(a) it does not get overheated

(b) it does not draw excessive current

(c) it can measure large potential differences

(d) it does not appreciably change the potential difference to be measured

56. A capacitor of capacitance 500 μF is connected to a battery through a 10 kΩ resistor. The charge stored on the capacitor in the first 5 s is larger than the charge stored in the next

(a) 5 s (b) 50 s

(c) 500 s (d) 500

57. When no current is through a conductor
    (a) the free electrons do not move
    (b) the average speed of a free electron over a large period of times is zero
    (c) the average thermal velocity of a free electron over a larger period of time is zero
    (d) the average of thermal velocities of all the free electron at an instant is zero

58. Two resistors A and B have resistances R_A and R_B respectively with $R_A < R_B$. The resistivity of their materials are ρ_A and ρ_B
    (a) $\rho_A > \rho_B$
    (b) $\rho_A = \rho_B$
    (c) $\rho_A < \rho_B$
    (d) the information is not sufficient to find the relation between ρ_A and ρ_B

59. Twelve equal resistances each of 2Ω form the edges of a cube. If a 2 volt battery having internal resistance 0.1 Ω is connected between two adjacent corners of a cube, then the current (in ampere) drawn from battery is
    (a) 60/53 (b) 30/19
    (c) 20/29 (d) none of the above

60. A steady current flows in a metallic conductor of non-uniform cross-section. The quantity/quantities constant along the length of the conductor is/are
    (a) current, electric field and drift velocity
    (b) drift speed only
    (c) current and drift speed
    (d) current only

61. A metallic resistor is connected across a battery. If the number of collisions of the free electrons with the lattice is somehow decreased in the resistor (for example, by cooling it), the current will
    (a) increase (b) decrease

(c) remain constant (d) become zero

62. A current i is flowing through a resistance connected to a primary cell. Now the resistance is reduced to half, the new current through the resistance is

(a) i (b) 2 i

(c) < 2 i (d) > 2 i

63. As the temperature of a conductor increases, its resistivity and conductivity change. The ratio of resistivity to conductivity

(a) increases

(b) decreases

(c) remains constant

(d) may increase or decrease depending on the actual temperature

64. If a wire of circular cross-section with radius r, free electrons travel with a drift velocity v when a current i flows through the wire. What is the current in another wire of half the radius and of the same material when the drift velocity is 2 v.

(a) i/4 (b) i/2

(c) i (d) 2 i

65. A current passes through a wire of nonuniform cross-section. Which of the following quantities are independent of the cross-section ?

(a) the charge crossing in a given time interval

(b) drift speed

(c) current density

(d) free-electron density

66. A resistor of resistance R is connected to an ideal battery. If the value of R is decreased, the power dissipated in the resistor will

(a) increase

(b) decrease

(c) remain unchanged

(d) none of the above

67. A current passes through a resistor. Let K_1 and K_2 represent the average kinetic energy of the conduction electrons and the metal ions respectively

    (a) $K_1 < K_2$

    (b) $K_1 = K_2$

    (c) $K_1 > K_2$

    (d) any of these three may occur

68. Two nonideal batteries are connected in series. Consider the following statements :

    (A) The equivalent emf is larger than either of the two emfs.

    (B) The equivalent internal resistance is smaller than either of the two internal resistances.

    (a) each of A and B is correct

    (b) A is correct but B is wrong

    (c) B is correct but A is wrong

    (d) each of A and B is wrong

69. As the temperature of a metallic resistor is increased, the product of its resistivity and conductivity

    (a) increases

    (b) decreases

    (c) remains constant

    (d) may increase or decrease

70. Two resistances R and 2R are connected in parallel in an electric circuit. The thermal energy developed in R and 2R are in the ration

    (a) 1 : 2 (b) 2 : 1

    (c) 1 : 4 (d) 4 : 1

71. A capacitor C_1 of capacitance 1mF and a capacitor C_2 of capacitance 2mF are separately charged by a common battery for a long time. The two capacitors are then separately discharged through equal resistors. Both the discharge circuits are connected at t = 0

    (a) the current in each of the two discharging circuits is zero at t = 0

(b) the currents in the two discharging circuits at t = 0 are equal but not zero

(c) the currents in the two discharging circuits at t = 0 are unequal

(d) C_1 loses 50% of its initial charge sooner than C_2 loses 50% of its initial charge

72. In an electric circuit containing a battery, the charge (assumed positive) inside the battery

(a) always goes from the positive terminal to the negative terminal

(b) may go from the positive terminal to the negative terminal

(c) always goes from the negative terminal to the positive terminal

(d) does not move

73. The net resistance of an ammeter should be small to ensure that

(a) it does not get overheated

(b) it does not draw excessive current

(c) it can measure large currents

(d) it does not appreciably change the current to be measured

74. Twelve equal wires, each of resistance 4 Ω are jointed to form a skeleton cube. The effective resistance of the cube between two points which are the diagonally opposite corners is

(a) 7/3 Ω (b) 28/5 Ω

(c) 10/3 Ω (d) 5/3 Ω

75. The product of resistivity and conductivity of a cylindrical conductor depends on

(a) temperature

(b) material

(c) area of cross-section

(d) none of these

76. Eleven equal wires each of resistance 4 Ω are jointed to form an incomplete cube. The total resistance between one end of vacant edge to the other end is

(a) 7/3 Ω (b) 28/5 Ω
(c) 10/3 Ω (d) 5/3 Ω

77. A current passes through a wire of non-uniform cross-section. Which of the following quantities are independent of the cross-section ?
    (a) free electron density
    (b) current density
    (c) drift speed
    (d) the charge crossing in a given time interval
78. When no current is passed through a conductor
    (a) the free electron do not move
    (b) the average speed of a free electron over a large period of time is zero
    (c) the average velocity of a free electron over a large period of time is zero
    (d) the average of the velocities of all the free electrons at an instant is zero
79. A uniform wire of resistance 50 Ω is cut into 5 equal parts. These parts are now connected in parallel. The equivalent resistance of the combination is
    (a) 2 Ω (b) 10 Ω
    (c) 250 Ω (d) 6250 Ω
80. In order to quadruple the resistance of a wire of uniform cross-section, a fraction of its length was uniformly stretched till the final length of the entire wire was 1.5 times the initial length. The fraction is
    (a) 1/2 (b) 1/4
    (c) 1/6 (d) 1/8
81. Two nonideal batteries are connected in parallel. Consider the following statements
    (A) The equivalent emf is smaller than either of the two emfs.
    (B) The equivalent internal resistance is smaller than either of the two internal resistances.

(a) both A and B are correct
(b) A is correct but B is wrong
(c) B is correct but A is wrong
(d) both A and B are wrong

82. The terminal potential difference of a battery exceeds its e.m.f. when it is connected
(a) in parallel with a battery of higher e.m.f.
(b) in series with a battery of lower e.m.f.
(c) in series with a battery of higher e.m.f.
(d) in parallel with a battery of lower e.m.f.

83. Mark out the correct options
(a) an ammeter should have small resistance
(b) an ammeter should have large resistance
(c) a voltmeter should have small resistance
(d) a voltmeter should have large resistance

84. The constants a and b for the pair silver-lead are 2.50 $\mu V/^{\circ}C$ and 0.12 $\mu V/(^{\circ}C)^2$ respectively. For a silver–lead thermocouple with colder junction at 0°C
(a) there will be no neutral temperature
(b) there will be no inversion temperature
(c) there will not be any thermo-emf even if the junctions are kept at different temperatures
(d) there will be no current in the thermocouple even if the junctions are kept at different temperatures

85. A d.c. voltage with appreciable ripple expressed as $V = V_1 + V_2 \cos \omega t$ is applied to a resistor R. The amount of heat generated per second is given by
(a) $\dfrac{V_1^2 + V_2^2}{2R}$ (b) $\dfrac{2V_1^2 + V_2^2}{2R}$
(c) $\dfrac{V_1^2 + 2V_2^2}{2R}$ (d) none of the above

86. Consider the following statements regarding a thermocouple

(A) The neutral temperature does not depend on the temperature of the cold junction.

(B) The inversion temperature does not depend on the temperature of the cold junction.

(a) both A and B are correct

(b) A is correct but B is wrong

(c) B is correct but A is wrong

(d) both A and B are wrong

87. How many lamps each of 50 W and 100 V can be connected in parallel across a 120 V battery of internal resistance 10 Ω, so that each glows to full power ?

(a) 2 (b) 4

(c) 6 (d) 8

88. An ammeter reads 0.90 A when connected in series with a silver voltameter. A steady current passed for 40 minutes deposits 2.60 gram of silver on cathode. If E.C.E. of silver is 1.118×10^{-3} g C^{-1}, then percentage error in the ammeter reading is

(a) 6.66 % (b) 7.3 %

(c) – 6.66% (d) – 7.3%

99. The electrochemical equivalent of a material depends on

(a) the nature of the material

(b) the current through the electrolyte containing the material

(c) the amount of charge passed through the electrolyte

(d) the amount of this material present in the electrolyte

90. The heat developed in a system is proportional to the current through it

(a) it cannot be Thomson heat

(b) it cannot be Peltier heat

(c) it cannot be joule heat

(d) it can be any of the three heats mentioned above

91. Which of the following statements is/are incorrect ?

(a) both peltier and joule effects are reversible

(b) both peltier and joule effects are irreversible

(c) joule effect is reversible, whereas Poltier effect is irreversible

(d) joule effect is irreversible, whereas poltier effect is reversible

92. Two resistors having equal resistances are joined in series and a current is passed through the combination. Neglect any variation in resistance as the temperature changes. In a given time interval

(a) equal amounts of thermal energy must be produced in the resistors

(b) unequal amounts of thermal energy may be produced

(c) the temperature must rise equally in the resistors

(d) the temperature may rise equally in the resistors

93. Which of the following statements is/are correct

(a) resistivity of electrolytes, decreases on increasing temperature

(b) resistance of mercury falls on decreasing its temperature

(c) when jointed in series a 40 W bulb glows more than a 60 W bulb

(d) resistance of 40 watt bulb is less than the resistance of 60 W bulb

94. The terminal potential difference of a battery exceeds its e.m.f. when it is connected

(a) in parallel with a battery of higher e.m.f.

(b) in series with a battery of lower e.m.f.

(c) in series with a battery of higher e.m.f.

(d) in parallel with a battery of lower e.m.f.

95. A silver and a copper voltameters are connected across a 6V battery of negligible resistance. If half an hour, 1 gram of copper and 2 gram of silver are deposited. The rate at which energy is supplied by the battery will approximately be (E.C.E. of copper = 3.294×10^{-4} g/C and E.C.E. of silver = 1.118×10^{-3} g/C)

(a) 16 watt (b) 32 watt

(c) 48 watt (d) 64 watt

96. An electrolysis experiment is stopped and the battery terminals are reversed
    (a) the electrolysis will stop
    (b) the rate of liberation of material at the electrodes will increase
    (c) the rate of liberation of material will remain the same
    (d) heat will be produced at a greater rate

97. The e.m.f. of a copper nickel thermocouple in μV per °C is given by $E = 16.34\ t - 0.02\ t^2$. The thermo electric power at 100°C in μV/°C/s will be
    (a) 12.5 (b) 12.34
    (c) 10.0 (d) 20.0

98. Faraday constant
    (a) depends on the amount of the electrolyte
    (b) depends on the current in the electrolyte
    (c) is a universal constant
    (d) depends on the amount of charge passed through the electrolyte

99. Two identical batteries each of e.m.f. 2 V and internal resistance 1Ω are available. The maximum rate of production of heat obtained in the external circuit is
    (a) 8 W (b) 4 W
    (c) 2 W (d) 1 W

100. Two resistors having equal resistances are joined in series and current passed through the combination. Neglect any variation in resistance as the temperature changes. In a given time interval
    (a) unequal amounts of thermal energy may be produced
    (b) equal amounts of thermal energy must be produced
    (c) the temperature may rise equally in the resistances
    (d) the temperature must rise equally at the resistors

101. An electric heating element consumers 500 W, when connected to a 100 V line. If the line voltage becomes 150 V, the power consumed will be
    (a) 500 W (b) 750 W

(c) 1000 W (d) 1125 W

102. Which of the following plots may represent the thermal energy produced in a resistor in a given time as a function of the electric current ? Represents the rate of production of thermal energy in the resistor

(a) 8 W (b) 4 W

(c) 2 W (d) 1 W

103. An electric kettle taking 3A at 200 V brings one litre of water from 20°C to the boiling point in 10 minute. Its efficiency is

(a) 33.3% (b) 66.6%

(c) 87.7% (d) 93.3%

104. A copper strip AB and an iron strip AC are joined at A. The junction A is maintained at 0°C and the free ends B and C are maintained at 100°C. There is a potential difference between

(a) the two ends of the copper strip

(b) the copper end the iron end at the junction

(c) the two ends of the iron strip

(d) the free ends B and C

105. A heater boils 1 kg of water in time t_1 and another heater boils the same water in time t_2. If both the heaters are connected in parallel, the combination will boil the water in time

(a) $\frac{t_1 t_2}{t_1 - t_2}$ (b) $\frac{t_1 t_2}{t_1 + t_2}$

(c) $\frac{t_1^2 t_2^2}{t_1 + t_2}$ (d) $\frac{t_1^2 - t_2^2}{(t_1 - t_2)}$

106. Fifty electric bulbs are connected in series across a 220 V supply and the illumination produced is I_1. Five bulbs are fused and remaining forty five bulbs are connected in series, the illumination produced is I_2. Which of the following is true ?

(a) $I_1 > I_2$ (b) $I_1 < I_2$

(c) $I_1 = I_2$ (d) none of the above

107. If the electric current in a lamp decreases by 5%, then the power output decreases by

(a) 2.5% (b) 5%

(c) 10% (d) 20%

108. A cell sends a current through a resistance R_1 for time t, next the same cell sends current through another resistance R_2 for the same time t. If the same amount of heat is developed in both the resistances, then the internal resistance of the cell is

(a) $(R_1 + R_2)/2$ (b) $(R_1 - R_2)/2$

(c) $R_1 R_2$ (d) $\sqrt{R_1 R_2}$

109. Consider a thermocouple made of iron and constantan. If the thermo e.m.f. of iron and constantan against platinum are + 800 and – 1700 μ V per 100°C difference of temperature, then the e.m.f. developed per °C difference of temperature between the junctions will be (in μV/°C)

(a) 25 (b) 50

(c) 75 (d) 100

110. Consider the following two statements

(A) Free-electron density is different in different metals.

(B) Free-electron density in a metal depends on temperature.

Seebeck effect is caused

(a) due to both A and B

(b) due to A but not due to B

(c) due to B but not due to A

(d) neither due to A nor due to B

111. Consider the statements A and B in the previous question. Peltier effect is caused

(a) due to both A and B

(b) due to A but not due to B

(c) due to B but not due to A

(d) neither due to A nor due to B

112. Consider the statements A and B in question 332. Thomson effect is caused

(a) due to both A and B

(b) due to A but not due to B

(c) due to B but not due to A

(d) neither due to A nor due to B

113. A house wiring supplied with a 220 V supply line is protected by a 6 ampere fuse. The maximum number of 60 W bulbs is parallel that can be turned on is

(a) 11 (b) 22

(c) 33 (d) 66

114. In the question 328, by what percentage the illumination of one bulb will change

(a) increase by 10%

(b) increase by 19%

(c) increase bt 23.5%

(d) increase by 24.5%

115. A thermocouple of resistance 1.6 Ω is connected in series with a galvanometer of 8.4 Ω resistance. The thermocouple develops an emf of 10μV per °C temperature difference between two junctions. When one junction is kept at 0°C and the other in a molten metal, the galvanometer reads 10 millivolt. The temperature of molten metal, when e.m.f. varies linearly with temperature difference, will be

(a) 1050°C (b) 1190°C

(c) 1275°C (d) 1545°C

116. A copper strip AB and an iron strip AC are joined at A. The junction A is maintained at 0°C and the free ends B and C are maintained at 100°C. There is a potential difference between

(a) the free ends B and C

(b) the two ends of copper strip

(c) the two ends of iron strip

(d) at the junction of copper end and iron end

117. In the above question no. 327, if the heaters are connected in series, the combination will boil the same water in time

(a) $\frac{t_1 t_2}{t_1 - t_2}$ (b) $\frac{t_1 t_2}{t_1 + t_2}$

(c) $\frac{t_1^2 - t_2^2}{t_1 + t_2}$ (d) $\frac{t_1^2 - t_2^2}{(t_1 - t_2)}$

118. In the question no. 328 by what percentage the total illumination will change

(a) increase by 10% (b) decrease by 10%

(c) increase by 11% (d) decrease by 11%

119. Masses of three wires are in the ratio of 1 : 3 : 5. Their lengths are in the ratio of 5 : 3 : 1. When connected in series with a battery the ratio of heats produced in them will be

(a) 1 : 3 : 5 (b) 5 : 3 : 1

(c) 1 : 15 : 125 (d) 125 : 15 : 1

120. A uniform wire connected across a supply produces heat H per second. If the wire is cut into three equal parts and all the parts are connected in parallel across the same supply, the heat produced per second will be

(a) H/9 (b) 9 H

(c) 3 H (d) H/3

121. Two similar head lamps are connected in parallel to each other. Together, they consume 48 W from a 6V battery, the resistance of each filament is

(a) 1.5 Ω (b) 3 Ω

(c) 4 Ω (d) 6 Ω

122. When a potential difference is applied across, the current passing through

(a) an insulator at 0 K is zero

(b) a semiconductor at 0 K is zero

(c) a metal at 0 K is finite

(d) a p-n diode at 300 K is finite if it is reverse biased

123. A wire of length L and 3 identical cells of negligible internal resistances are connected in series. Due to the current, the temperature of the wire is raised by ΔT in a time. A number N of similar cells is now connected in series with a wire of the same material and cross-section but of length 2L. The temperature of

the wire is raised by the same amount ΔT in the same time. The value of N is

(a) 4 (b) 6

(c) 8 (d) 9

124. A short-circuited coil is placed in a time-varying magnetic field. Electrical power is dissipated due to the current induced in the coil. If the number of turns were to be quadrupled and the wire radius halved, the electrical power dissipated would be

(a) halved (b) the same

(c) doubled (d) quadrupled

Fill in the Blanks

1. .... electrons flowing per sec constitute a current of one ampere.
2. 1 Faraday is the charge required to deposit .... of any substance.
3. Charge required to deposit 9 gms of aluminium is equal to ....
4. When same charge is passed through different voltameters connected in series, then masses of different substances liberated are found to be in the direct ratio of their ....
5. The drift velocity of free electrons is of the order of ....
7. In metallic conductor, electrons are not accelerated continuously but for a very .... interval of time, called as ....
6. Temperature coefficient of resistance for .... is positive but for .... and .... it is negative.
7. Among the bulbs of different wattages joined in parallel, the bulb of .... wattage glows maximum, but in series the bulb of .... wattage glows maximum.
8. Alloys like constantan, nichrome etc. have high .... and low ....
9. When temperature is increased, the resistance of metallic conductor .... but that of semiconductor

10. A constant potential difference is applied across a metallic conductor. The heat produced in it is ....

    if both the length and the radius of the wire are doubled.
11. 1 KWH is equal to .... Joules.
12. 5 bulbs of 100 watt each glow 10 hours every day for 30 days. The electric is KWH consumed will be ....
13. Resistance of a metallic conductor is .... proportional to relaxation time.
14. With rise in temperature, the relaxation time gets ....
15. Heat produced in a given metallic conductor is directly proportional to .... of the current.
16. Resistance of a highest wattage bulb is ....
17. Maximum current that can be passed through a bulb is .... proportional to its power.
18. Ohm's law holds good in metallic wires for .... only.
19. V-i graph is .... for circuit components like diode valve, triode valve, transistor etc.
20. The circuit components, which do not obey ohm's law even for small currents, are called as ....
21. Resistivity of electrolyte .... with rise in temperature.
22. Resistivity of semiconductors .... with rise in temperature.
23. Resistance of a diode valve is .... at difference values of applied potential difference.
24. Charge carries responsible for conduction of electricity in electrolytes are ....
25. In a conductor free electrons always move from .... to ....

True/False

1. A copper wire of length L and cross sectional area A carries a current I. If the specific resistance of copper is S, the electric field in the wire is IS/A.

2. The drift velocity of electrons in a metallic wire will decrease if the temperature of the wire is increased.
3. Two wires, each of radius r, but of different materials are connected together end to end. If the densities of charge carriers in the two wires are in the ratio 1 : 4, the drift velocity of electrons in the two wires will be in the ratio 1 : 2.
4. A constant voltage is applied between the two ends of uniform metallic wire. If both the length and radius of the wire are doubled, then heat developed in the wire will remain same.
5. The drift velocity of an electron is doubled if the applied electric field across the conductor is doubled.
6. The temperature coefficient of resistance of a wire is 0.00125/°C. The resistance of the wire is 1 ohm at 300°K. The resistance will be 2 ohm at 1100°K.
7. Kilo-watt hour is the unit of electric power.
8. A 200 W bulb has more resistance than a 100 W bulb.
9. Heat energy produced in a given current carrying wire is directly proportional to the square of the current.
10. Heater wire must have high resistance and high melting point.
11. Fuse wire must have high resistance and low melting point.
12. In a chain of bulbs, 50 bulbs are joined in series. One bulb is fused now. If the remaining 49 bulbs are again connected in series across the same supply then light gets decreased in the room.
13. Charge required to deposit 1 gm-equivalent of any substance is 9.65×10^7 coulombs.
14. If 25 watt and 100 watt bulbs are joined in series and connected to mains, then 100 watt bulb will glow brighter.
15. Two electric bulbs of 50 and 100 watt are given. When connected in series 50 watt bulb glows more but when connected in parallel 100 watt bulb glows more.
16. Two bulbs of same wattage, one having a carbon filament and the other having a metallic filament, are connected in series. Metallic filament bulb will glow more brightly than carbon filament bulb.

17. It is advantageous to transmit electric power at high voltages.
18. Water boils in an electric kettle in 15 minutes after being switched on. For boiling the water in 10 minutes, using the same mains supply, length of heating element must be increased.
19. Current is passed through a metallic wire heating it red. When cold water is poured on half of its portion, then rest of the half portion become still more hot.
20. An electric bulb is first connected to a D.C. source and then to an A.C. source having the same voltage. Bulb glows with same brightness in both the cases.
21. Though the same current flows through the line wires and the filament of the bulb but heat produced in the filament is much higher than that in line wires.
22. A heater joined in series with a 50 watt bulb is connected to mains. When 50 watt bulb is replaced by 100 watt bulb, heater gives less heat.
23. When the ends of a wire are connected to a battery, the current is slightly higher initially and becomes steady at a lower value after some time.
24. Three wires of resistances 2, 3 and 4 ohms are given. A constant potential difference is applied across each wire one by one. Rate of production of heat will be maximum in 4 ohm resistance.
25. When the radius of a copper wire is doubled, its specific resistance gets increased.
26. Resistance of an Eureka wire gets doubled when its radius is halved and length is reduced to one fourth of its original length.
27. A current flows in a conductor only when there is an electric field within the conductor.
28. When a wire is stretched to make its length doubled, its resistance gets increased.
29. A current is flowing in a copper wire. When same current is allowed to flow in another copper wire of double radius, then drift velocity gets increased.
30. In above question, when same current is passed through iron wire of same thickness, then drift velocity is less in comparison to that in copper wire.